E-commerce Usability

Essential readings from Taylor & Francis:

Designing Usable Electronic Text
Andrew Dillon, University of Texas, USA
ISBN 0–7484–0112–1 (hb)
ISBN 0–7484–0113–X (pb)

Inclusive Design Guidelines for Human–Computer Interaction
Edited by Colette Nicolle, HUSAT, UK and Julio Abascal, University of the Basque Country, Spain
ISBN 0–7484–0948–3 (hb)

User Interface Design for Electronic Appliances
Edited by Konrad Baumann, Philips Consumer Communications, Vienna, Austria and Bruce Thomas, Philips Design, Vienna, Austria
ISBN 0–415–24335–1 (hb)

Information and ordering details
For price availability and ordering visit our website
www.ergonomicsarena.com
Alternatively our books are available from all good bookshops

E-commerce Usability

Tools and techniques to perfect the on-line experience

David Travis

Taylor & Francis
Taylor & Francis Group
LONDON AND NEW YORK

First published 2003
by Taylor & Francis
11 New Fetter Lane, London EC4P 4EE

Simultaneously published in the USA and Canada
by Taylor & Francis
29 West 35th Street, New York, NY 10001

Taylor & Francis is an imprint of the Taylor & Francis Group

Typeset in Sabon by
HWA Text and Data Management Ltd, Tunbridge Wells
Printed and bound in Great Britain by
TJ International Ltd, Padstow, Cornwall

British Library Cataloguing in Publication Data
A catalogue record for this book is available from the British Library

Library of Congress Cataloging in Publication Data
A catalog record for this book has been requested

ISBN 0–415–25834–0

For Joshua, Bethany, Sam
and Gret

Contents

List of figures

List of tables

Preface

We are all familiar with airport bookstall best sellers that promise fame, fortune and success in relationships. Many of them – at least the ones I have skimmed whilst waiting for one of the world's favourite airlines – are deeply disappointing. Not because they do not work (to be honest I have not really been able to try them out fully) but because they tend to miss out on the tricky bit in the process. For example, a book on how to become really rich will often contain such helpful nuggets as 'make lots of money'. If you then turn to the *How to Make Lots of Money* book, the advice is to start a successful business. The *How to Start a Successful Business* book then suggests that the key is to come up with a simple idea that everyone wants. The *How to Come up with Clever Ideas* book typically talks about brainstorming and all manner of 'creative ideas'.

Where they fail is that the ideas themselves are often fairly simple to generate; it's turning them into real solutions that involves a lot of hard work.

One of the widely quoted examples of the 'ah-ha' principle at work concerns Kekulé von Stradonitz in 1865 who was struggling to interpret his data on the structure of benzene. Allegedly, after a particularly gruelling day in the laboratory, he fell asleep and dreamed of a group of six snakes. At some point, these dancing snakes each caught another's tail in its mouth, in such a way that they formed a six-snake ring. When Kekulé awoke, he had stumbled on the resonating ring structure, which made sense of all his work. Now I doubt Kekulé was the first to dream of snakes (indeed I believe they are a rather well-known Freudian symbol) but he was the first one to make the link. The idea on its own meant little, but combined with serious groundwork, provided the apparently trivial spark which made sense of it all.

Clever ideas in computing abound, especially on the web. It is converting them into a sound business model which is more difficult and usually causes most would-be 'Bill Gates' to fall by the wayside. E-commerce is an area where bright ideas are common but sound business models are not.

One of the areas where I would claim some expertise (on the basis that I have been working in the area for some big names for many years) is the ergonomics and psychology of retailing. I am always both amused and astounded by the way many e-commerce sites seem to understand customer psychology and then either ignore it or do the opposite. Real stores design their layouts in such a way that customers are enticed and led through product offerings. High visibility locations (for example, at the ends of aisles) are used carefully to promote particular products and the store entrance is prize territory. Many stores have lots of doors – often automatic. These are not to allow customers out in the event of a fire, but to make it easy for them to enter. Staff are trained to welcome customers, maintain eye contact and generally make them feel welcome.

Yet if we are to believe many e-commerce sites, customers should be challenged at the door to give their address before guessing where merchandise is hidden. Slow response, unhelpful graphics, hidden functionality and unreliable links test the customer's stamina to the limit – before they have even purchased anything. Feedback is often incomplete or even missing.

If it were a real store, customers would walk next door or even travel further afield. On a web site this takes no more than a click or two.

So if usability is so obviously important in e-commerce, why is it not taken more seriously? I would argue that there are three main reasons:

- many people believe that since the web is new, we need to start again learning how to make it usable;
- others argue that the web is unique so nothing that we knew about conventional usability applies.

But probably the most significant reason is that:

- it is difficult to know exactly what to do to ensure good usability.

There are lots of 'gurus' with helpful advice like 'do what I say' who are great at publicising usability but who are not so good at passing on their skills to regular web site designers.

Putting usability knowledge and skills in the public domain is one of the prime motivations of our work in the International Organisation for Standardisation (ISO) Ergonomics of Human System Interaction Committee. Since 1983 we have developed a number of standards aimed at improving the usability of computer hardware and software. The multipart ISO 9241 is one of the better known standards and contains one part (Part 11) specifically devoted to 'Guidance on Usability'. Our experience in developing these standards made us realise that in the computer industry we would always be 'playing catch-up' trying to keep up with the fast pace of hardware and software development.

In the mid-1990s we therefore decided to develop a process standard which would be technology-independent and which would represent international consensus on best practice. In 1999, ISO 13407 'Human Centred Design Processes for Interactive Systems' was published with the explicit aim of providing project managers with a means of ensuring that their design processes have a high likelihood of developing usable systems.

In a sense, it is like the airport best sellers I was criticising earlier. If the process is followed completely, then the designer is assured of a usable product. Of course, there is a catch. Success is only guaranteed if appropriate criteria are identified initially and if the process iterates until they are achieved. It may not be possible to complete the process if the original design objective is unachievable. Designers may not be able to solve some of the problems which testing with real users provides. But the key point is that the standard provides an internationally agreed framework for the usability process.

Since its publication in 2000, it has been well received in Japan (where it is seen as relevant to the design of a wide range of products and systems). In the USA it was one of the key stimuli (along with ISO 9241-11: Guidance on Usability) for the development of the ANSI/NCITS Common Industry Format for reporting usability results. In the UK, it has been promoted as a method for helping public sector systems achieve the Government's objective of effectiveness of IT systems.

In System Concepts, we use this framework to inform all our usability work from mobile telephones to government information systems, from computer printers to interactive television and, of course, to web sites.

David Travis has drawn together our experiences with this process in an informative and compelling style. In this book, he has tailored the customer-centred design process to the difficult, but not impossible problem, of making e-commerce sites that people can and do use successfully. Standards – particularly international standards – are written in a peculiar form of English, which often sounds stilted. One of David's great strengths is that he has written a book which is not only a highly *effective* and an *efficient* way of communicating about e-commerce usability, he has also made it *satisfying*, sometimes even fun, to read. In other words, it is a very usable book about usability. The book is rich in informative examples and even the laziest reader should find inspiration from the practical experiences which fill each page.

I was delighted when David asked me to write this preface and I am pleased to commend this book wholeheartedly. I am sure that designers too will find it an invaluable resource and I look forward to the improvement in usable web sites that should result.

Tom Stewart
Managing Director, System Concepts
Chairman, ISO TC 159 SC4 Ergonomics of Human System Interaction

Acknowledgements

This book started life as a training course. The course content was knocked into shape by dozens of clients; and hundreds of delegates have usability tested the ideas and helped fix those that were confusing, ambiguous or difficult to apply. Amongst these, I would particularly like to thank Leigh Davies at Thomas Cook who helped structure the initial course.

But a course means nothing if the theory cannot be put into practice. In my consulting work, I have been fortunate to work with clients on some truly challenging projects. Indeed, all of the practical ideas in this book arose from usability assignments with clients. There are too many to thank individually but Debbie Mrazek at HP stands out for her ability to challenge and coach in equal measure.

And none of *that* would have been possible if Tom Stewart hadn't invited me to the Savoy for a Christmas party many years ago and encouraged my interest in standards and consultancy. Thanks Tom!

A number of people supported and encouraged me in the writing of this book and provided comments on the content and structure. In particular, I would like to thank Nigel Bevan, Marty Carroll, Jeff Johnson, Deborah Mayhew, Debbie Mrazek, Keith Instone, John Rhodes, Ben Shneiderman and Tom Stewart. Nick Freeman of manha.com did a great job on the figures; and Sarah Kramer at Taylor & Francis motivated me to deliver the book, rather than re-write it yet again. I want to thank all of them for their time and point out that the remaining errors are, of course, mine.

Finally: thanks Gret. Can I come out now?

1 Introduction

There are hundreds of books aimed at the people who design web sites. Books that tell you how to write HTML with the very latest editing tools. Books that show you how to design 3D buttons in Adobe PhotoShop. Books that explain how to code in Perl and Java. Books that discuss fashionable mark-up languages, such as XML, DHTML and VRML.

This book takes a different approach: it assumes that the people who *use* web sites just want an easy life.

This assumption states that people will always choose a simple way of achieving their goals over a complex way. So this book on web site development hardly mentions technology. Instead, it focuses on the *customers* of the technology: it explains how to design e-commerce sites that ordinary people can use.

Such an approach is sorely needed. There is hardly an area of work or business life that has not been affected by computing technology. But customers of this technology are almost universally ignored. Operating systems crash on a daily basis and applications bail out with an indecipherable error message. Simple and obvious tasks are implemented in ways so convoluted that customers often need to be taught how to carry them out. New versions of software are released that look different – they have more intricate icons, different beeps and swish animated features – but the fundamental problems remain (and new problems are introduced). Is it a coincidence that the only other industry that refers to its customers as 'users' (and treats customers with equal contempt) is the drug industry?

E-commerce is the latest development that attempts to persuade customers that technology will make their lives easier. Simple observation lends the lie to this assertion. In offices and homes throughout the world, customers of e-commerce sites are suffering from what I like to call 'technological Tourette's syndrome'. Sufferers of *real* Tourette's syndrome have a compulsion to swear, twitch and shout. And indeed, customers of e-commerce sites often act the same way.

'My "cookie expired"? What the **** does that mean?'

'****! How do I actually buy this thing!'

'I don't believe it! This ******* page crashed my browser!'

We see that people who use e-commerce sites grunt, swear and jerk, just like people with real Tourette's syndrome.

Yet the notion of usability is not new. It was just ahead of its time. Only a few years ago, product manufacturers disregarded usability because the benefits accrue to the people who buy a product, not the people who make it. For example, for a manufacturing company to invest funds in improving the usability of a VCR, the product manager needed to be convinced that consumers base their purchase decision on usability. But everyone knows that if we ask a customer to choose between two VCRs, the customer's decision is based mainly on features, price and aesthetics. It is only later – when the customer cannot work out how to stop the clock from flashing '12:00' – that usability (or the lack of it) becomes a motivating factor. And by then it is too late.

We can conclude that manufacturers are rarely bothered with optimising usability because they make money from the one-off sale of the product. The costs of poor usability are borne entirely by the 'user'.

The contempt is almost palpable.

Convergence has tipped this business model on its head. Many products nowadays only generate revenue for companies if they are used. Indeed, some products, such as mobile phones and the set-top box for interactive television, are sold to consumers at a fraction of their manufacturing price – sometimes given away. High-tech companies now make money from people *using* the product, not buying it: just as razor manufacturers before them discovered that the real market was in the blades, not the razor. This applies in spades for e-commerce sites: if a customer cannot use your site, they will not use your site, no matter how much is spent on advertising and marketing. Conversely, the easier it is for customers to buy, the more customers will buy. In this model, improved usability has short- and long-term benefits for both the company and the individual product manager.

Almost on a monthly basis independent surveys bear this assertion out, highlighting the amount of business lost by sites that are difficult to use. We read that people who want to buy products are unable to because of navigation difficulties: customers are unable to find the correct page to choose the product, or are unable to find the payment option. We read that sites crash, are under construction, or are otherwise inaccessible.

So it is now *obvious* that web sites need to be usable. The good news is that usability has finally come of age.

The bad news is that usability is perceived as screen design: choosing the correct fonts, colours and icons. In fact, usability is a process: it is not something that can be stapled on at the end of development. To say that usability is about screen design is as erroneous as saying that branding is all about a good logo. Of course screen design plays a role in usability, but it is a small role. This means that optimising the colours, fonts and icons on your site will improve usability by, at best, 15 per cent. It's like the old adage: you can put make-up on a pig, but it's still a pig.

Screen design is just one of three important components. Web sites that score high on usability also show a second key feature: consistency. Consistency accounts for about 25 per cent of a web site's usability. We can all point to annoying inconsistencies in (or between) much of the software we use. For example, the 'Cut and paste' operation in Microsoft Excel works differently from every other piece of software – even other Microsoft products. Choose 'Cut' in Microsoft Word and the selection disappears. Choose 'Cut' in Microsoft Excel and 'marching ants' appear around the selection, but it remains where it is. In this example, the inconsistency causes no more than a hesitation ('Did I choose "Cut" or something else?'), but move the application domain to a nuclear power station or the control room of a chemical plant and one man's hesitation quickly becomes another man's environmental catastrophe.

The third component of usability, the remaining 60 per cent, is accounted for by task focus.

You know a web site has task focus when you get a warm feeling that the person who designed the site knew exactly what you wanted to do. The site works the way you expect. There is no need to go searching through menus or dialogue boxes for obscure commands: the main things you want to do are there in front of you – easy to find and simple to carry out. The web site delights you. Another example comes from the world of successful computer games: very quickly, the 'interface' disappears and you are exploring the world of the game – the task.

Rules for good visual design are plentiful – you can get them from a book. Using code libraries and then testing against a style guide can achieve consistency. But achieving task focus is much more complicated – it requires a process, and it is the substance of this book.

The process described in this book for ensuring usability is based closely on a recent usability standard, *Human-Centred Design Processes for Interactive Systems* (BS-EN ISO 13407: 1999). One of the strengths of this standard is that it can be tailored to support existing design methodologies. Because the process is not tied to any one notation or tool set, it can be easily adapted to create models in whichever notation and tool set the programming team uses. By following the human-centred design process described in the standard, project managers, designers and developers can ensure that systems will be effective, efficient and satisfying for customers.

The standard describes four principles of human-centred design:

1 active involvement of customers (or those who speak for them);
2 appropriate allocation of function (making sure human skill is used properly);
3 iteration of design solutions (therefore allow time in project planning);
4 multi-disciplinary design (but beware overly large design teams).

... And four key human-centred design activities:

1 understand and specify the context of use (make it explicit – avoid assuming it is obvious);
2 specify user and socio-cultural requirements (note there will be a variety of different viewpoints and individuality);
3 produce design solutions (note plural, multiple designs encourage creativity);
4 evaluate designs against requirements (involves real customer testing not just convincing demonstrations).

The standard itself is generic and can be applied to any system or product. This book tailors the process to e-commerce design based on many years' practical experience. The aim is to describe a practical method to help software developers, project managers, business analysts and user interface designers build better e-commerce sites. The approach is iterative and deliverable-based; forms or designs are completed at the end of each stage, which then constitutes the requirements and design specification for a project.

The book does not assume any background in human factors and usability: its aim is to provide step-by-step guidance to help non-experts carry out each stage of a proven customer-centred design process. The book is based on practical consultancy assignments with a number of clients in the financial and high-technology sectors (including Hewlett-Packard, Telewest, Motorola, *The Financial Times*, Thomas Cook and Philips).

The customer-centred design process has four steps (see Figure 1.1).

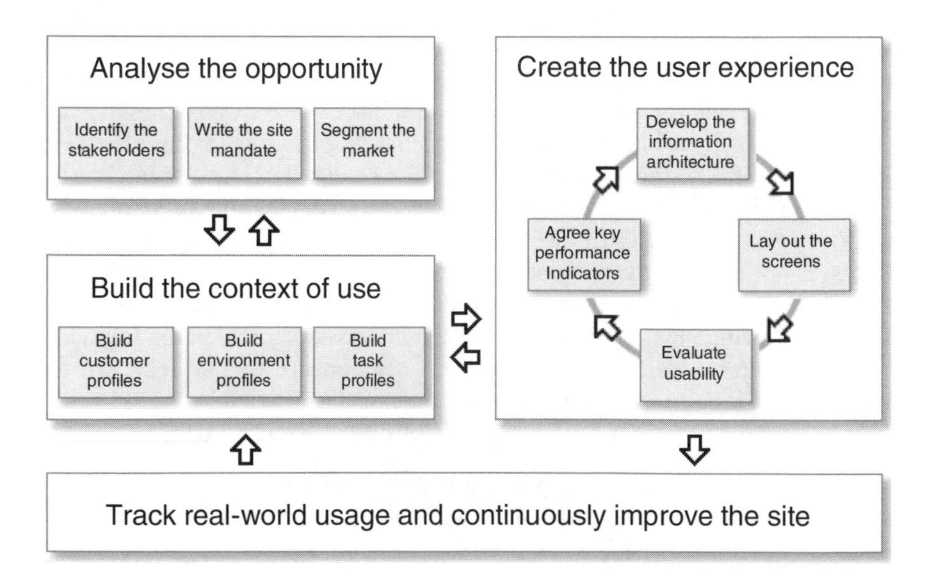

Figure 1.1 The customer-centred design model

A good way to start is with an assessment of your current design method, so I recommend you take the quick test in Chapter 2. This will show the strengths of your current approach and should identify areas for improvement.

Next, turn to the first part of the book that describes how to *analyse the opportunity*. This stage provides the business context for the web site: within this stage we will identify the stakeholders (Chapter 3), identify why the site is being developed (Chapter 4), and segment the market for the site (Chapter 5).

The second part of the book describes how to *build the context of use*. This results in a rich description of customers (Chapter 6), the environment in which they access the site (Chapter 7) and a description of realistic activities or 'task scenarios' (Chapter 8).

Part three shows how to use these data to *create the user experience*. This is an iterative process. Chapter 9 starts the process by showing how to develop key performance indicators for the web site: quantitative measures, based on key customer and business requirements, that the management team use to determine if the site is ready to 'go live'. Chapter 10 describes practical techniques to develop the information architecture (the high-level, conceptual model), with Chapter 11 devoted to laying out the screens (the detailed design), starting with paper sketches and then moving to electronic slide shows or interactive prototypes. Chapter 12 shows how to evaluate the site's usability, by using both usability experts and representative customers.

The final step in the process, described in the last part of the book, is to *track real-world usage and continuously improve the site* once it has been deployed.

This book aims to present a concise summary of each of these areas. Readers who are keen to learn more will hopefully find the concluding sections of each chapter, 'Further reading and weblinks', useful. To help you decide if a weblink is truly useful before firing up your browser, I have tried to annotate most of them with direct quotations from the article. The final chapter of the book provides some practical ideas on ways to apply the techniques immediately.

There are four key benefits from a customer-centred approach: higher revenues, loyal customers, improved brand value and process improvement.

- **Higher revenues:**
  - fewer changes downstream means earlier time to market;
  - earlier time to market brings competitive advantage;
  - customers use all of the site's functionality, not just a sub-set;
  - early and continuous customer involvement reduces lifecycle costs;
  - customers cost less to service (they won't need to phone up to check their order went through).

- **Loyal customers:**
  - customers remain loyal – loyal customers generate repeat business, demonstrate immunity to the competition, provide higher margins and are less price sensitive;

  - value to customers is delivered in the first release of the site as well as upgrades;
  - free word-of-mouth exposure.

- **Improved brand value:**
  - customers learn more quickly how to use the site;
  - improved usability provides a competitive edge;
  - higher service quality leads to improved customer satisfaction;
  - customers can focus on their goals rather than the web site: this leads to increased productivity and fewer errors.

- **Process improvement:**
  - reduced rework to meet customer requirements: 80 per cent of software re-writes are due to important functionality being missed the first time;
  - the process keeps developers focused on important business metrics, such as conversion rate;
  - development, marketing and external contractors improve communication and can better orchestrate their efforts;
  - risks are managed and reduced by helping you prioritise features and product offerings.

Read on to see how you can realise these benefits on your own project, and don't forget to check out usabilitybook.com for links, updates and free downloads related to the book.

Further reading and weblinks

Books and articles

Mayhew, D.J. (1999) *The Usability Engineering Lifecycle*. San Francisco, CA: Morgan Kaufman Publishers.

Nielsen, J. (1999) *Designing Web Usability: The Practice of Simplicity*. Indianapolis, IN: New Riders Publishing.

Shneiderman, B. (1997) *Designing the User Interface: Strategies for Effective Human–Computer Interaction*, 3rd edition. Reading, MA: Addison-Wesley.

Web pages

Travis, D.S. (2002) 'E-commerce usability'. http://www.usabilitybook.com. This web site accompanies the book. Check here for updates to the text, free downloads, and links to useful resources.

'Usable Web'. http://www.usableweb.com. Usable Web is a well organised collection of links to articles on usability and the web (such as information architecture, user interface issues, and design).

'Useit'. http://www.useit.com. Useit presents Jakob Nielsen's bi-monthly articles on web design and usability. The articles are passionate, outspoken and always worth reading.

'WebWord'. http://www.webword.com. Webword is a well-managed, up-to-the-minute, collection of links to articles and news items on usability.

2 Is your site customer centred?

You may be starting out on a new project, or perhaps this is a re-design of an existing site. In either case, this chapter contains a short test that you can use to measure the likely success of your site. The questions assess how closely you followed (or plan to follow) a customer-centred design approach.

You can use the results of this test as a 'before' and 'after' measure for your site, or you can use it to identify particular areas for improvement. An electronic version of this test is available from the book's web site, www.usabilitybook.com. You can also use it as a checklist after you complete each phase of the design to check that the key issues have been addressed.

Answer each question using a scale of 0–2, using the following key:

0: No
1: Kind of
2: Yes

Analyse the opportunity

1 Have all the project stakeholders been identified (not just the end users)? ______
2 Have the motivations of each stakeholder been made explicit? ______
3 Has the list of stakeholders been prioritised and is there an explicit management strategy for the most important stakeholders? ______
4 Does the project have a site mandate or vision statement that clearly communicates the aims of the project to the development team? ______
5 Is it clear how the site will make money? ______
6 Are the objectives of the site clear and unambiguous, and are they consistent with the organisation's overall business goals? ______
7 Is the value that customers will get from using the site clear and obvious? ______

8 Have competitor web sites been identified? ______
9 Has customer research been carried out to identify the most appropriate market or beachhead segment that will be the prime focus of the first release? ______
10 Are the target customers clearly identifiable (i.e. the site avoids trying to be 'something for everybody')? ______

Build the context of use

1 Have potential customers been consulted and have their goals been identified? ______
2 Is there a specification of the range of intended customers for the site? ______
3 Is there a specification of the environments (physical, socio-cultural and technical) within which customers will use the site? ______
4 Is there a specification of the different tasks that customers will want to carry out at the site? ______
5 Has this list of customer tasks been prioritised and have the critical and frequent tasks been identified? ______
6 Has this prioritised list of tasks been used to deliver the web site in stages, with the most important functionality delivered first? ______
7 Was information about customers, their tasks and their environments collected using a range of techniques (e.g. surveys, interviews, observation)? ______
8 Has the context of use information been reviewed and verified? ______
9 Has the context of use information been presented to the design team in an engaging and accessible way (e.g. as personas and scenarios)? ______
10 Has the context of use information been used to drive the design process? ______

Create the user experience

1 Have key performance indicators been set and has the site been evaluated against these metrics? ______
2 Are these criteria based on clear customer input and do they cover the areas of customer performance (effectiveness and efficiency) as well as customer satisfaction? ______
3 Has the conceptual model for the site been identified? ______
4 Has a content inventory been produced, and has this been sorted, ordered and categorised to match customers' mental model of the information? ______

5 Is the navigational framework of the site and the terminology used for navigation items based on this customer research? ______
6 Have various prototypes been developed and evaluated and have the results of these evaluations been used to improve the designs? ______
7 Has the design been tested for usability with representative customers? ______
8 Has the design been changed to address the findings of these evaluations? ______
9 Have usability defects been prioritised based on their impact on customers and have the defects been tracked to completion? ______
10 Has the design been documented in a style guide? ______

Track real-world usage and continuously improve the site

1 Is feedback from customers (e.g. help desks, post-release surveys and customer visits) collected regularly and frequently to ensure that the site continues to meet business and customer needs? ______
2 Is this information used to identify changes in the customer base, their environments and their tasks? ______
3 Is the specification of the context of use (customer, environment and task profiles) updated regularly to reflect any major changes? ______
4 Is this information used to identify the key areas of the site to maintain and enhance? ______
5 Are measures of conversion rate, fulfilment and customer retention tracked regularly and frequently? ______
6 Has there been verification that the business requirements were met? ______
7 Has there been verification that the stakeholder requirements were met? ______
8 Has there been verification that the site objectives were met? ______
9 Has there been verification that the key performance indicators were met? ______
10 Did the design team hold a post-implementation meeting to discuss the effectiveness of usability processes and identify areas of improvement? ______

Interpreting your results

Add up your scores for each section independently. This provides you with an index score for each step in the customer centred design process. Plot your scores on the 'Before' template below by shading in the segments on the relative quadrants of the diagram.

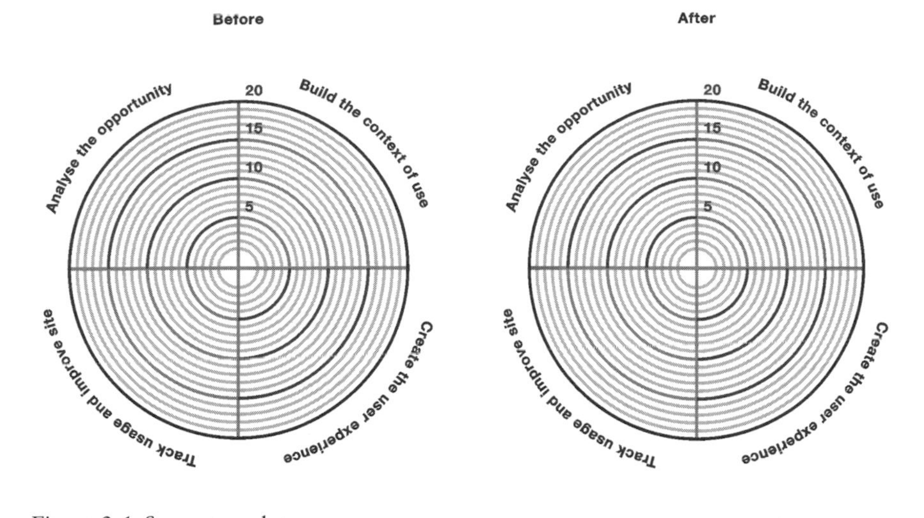

Figure 2.1 Score templates

Plotting your scores in this way helps identify potential areas for improvement. For example, you may find that you have scored high on 'Analyse the opportunity' but low on 'Build the context of use'. In that case, you should make sure you read Part II of the book.

You should also calculate your overall score by adding up the scores for each section. You can interpret this score in the following way:

65–80: Excellent. You apply usability principles to the design of e-commerce sites and your company has integrated usability fully into every phase of the lifecycle. Customers are involved in the design process early and often.

45–64: Good. You focus on usability during some parts of the development lifecycle and try to match the site to the needs of customers. You want to involve customers more, but your company feels that customer involvement isn't necessary at certain lifecycle phases (probably the pre- or post-design phases).

25–44: Fair. Your company recognises that usability is important, but there is no focus for it in the organisation. Usability methods are applied in an *ad hoc* manner. On those occasions when customers are involved, it is usually too late to make many changes.

0–24: Poor. Your company considers usability to be unimportant. Customers are rarely, if ever, asked to evaluate site designs. Your e-commerce development is unlikely to ever turn a profit.

Now you have some idea for potential areas of improvement. Read on, apply the techniques that you learn and then return to this test to check that your improvement continues.

Part I

Step 1

Analyse the opportunity

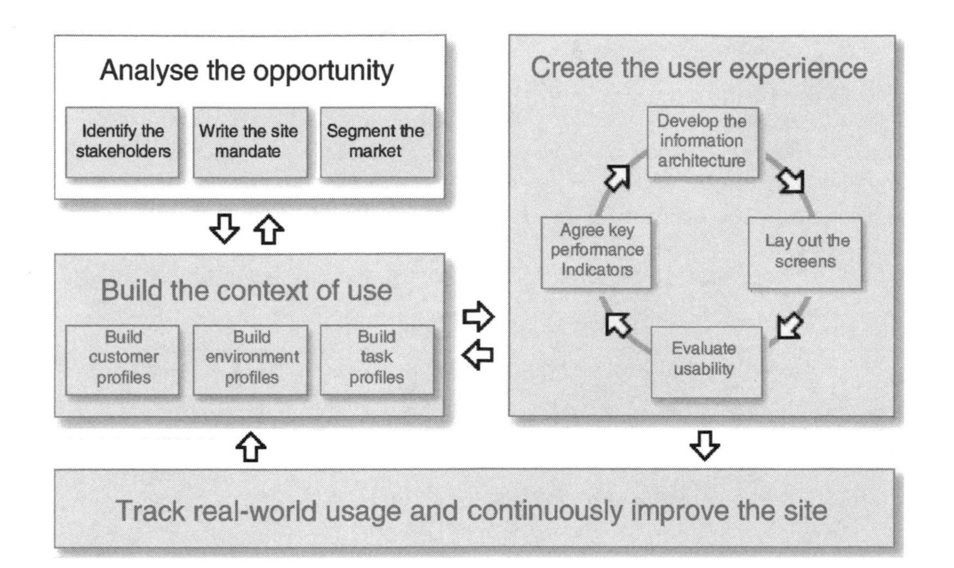

3 Identify the stakeholders

When designing an e-commerce site, it is tempting to ignore the customers of the site (who can sometimes be hard to define) and focus instead on the functionality (which is usually well specified). One benefit of this is that coding and prototyping can start very early, often within the first few days of a project. (Indeed, in some projects the prototype pre-dates the project – it is used to gain funding.) But there are costs with this approach: most notably, the project has spent inadequate time on capturing requirements, and once a prototype is in circulation it becomes very difficult to change. The prototype *becomes* the specification.

Even projects that identify key customers early on in the project often tend to identify just one or two people. Common examples include the person who is paying you to do the work (the client or the venture capitalist) and perhaps an 'end user' who at this point may be quite vague and elastic.

But many people will have an influence on the success of your web site – including a number of people who may never actually use it.

Perhaps the most familiar example is the CEO of the organisation who sees the web site as the company's 'face' to the world. He or she may place particular demands on the design and expect to see certain features that are of no interest to your real customers. These 'boss' features as we might call them could compromise the usability of the site. But excluding them may prevent the design from being accepted by the client. How do you manage this situation?

An analysis of all the people who will come into contact with your site – the stakeholders – will help you address this important first step. Stakeholder analysis is a common procedure in project management and is a vital early step in the customer centred design process. It will help flush out all of the important roles and responsibilities and identify people that can make the project succeed or fail.

Stakeholder analysis has four components:

- identify the stakeholders;
- uncover stakeholder motivations;

- prioritise the list;
- devise a management strategy.

We now address each of these steps in turn.

Identify the stakeholders

Our first step is to identify the full range of people who will interact with the web site, no matter how superficially. This includes all those people who have an interest in the success or failure of the site. Typical roles are shown in Table 3.1.

Go through each of these roles for your own project. Some of the role titles may be different on your project, and some roles may not exist at all. But for those that do, write down names of individuals: specific people who you can contact.

Once you have gone through this list, you need to check that you have coverage. To do this, identify people or businesses that will make money or save time by using the site; and identify the job titles of people who will directly interact with the web site to achieve those benefits.

You should now have a list of stakeholder names. We now need to think about each stakeholder's motivations: what is that person's interest in this project?

Uncover stakeholder motivations

Once we have a list of names, we can begin to uncover the motivations of each stakeholder. Human motivation is a complex topic and the aim here is simply to uncover motivations as they apply to our particular e-commerce development. The purpose of assigning names to roles in the previous step was to make sure that we start to think of individuals as real people (and not as stereotypes) in this step. People have beliefs, doubts, interests, values, feelings, desires and worries and all of these can have an impact on their attitudes to a new development.

As an example, Table 3.2 shows a list of possible concerns for each of the stakeholder roles identified below. This list should be taken merely as a starting

Table 3.1 Typical stakeholder roles

- Clients or sponsors
- Customers or users (57 varieties)
- Shareholders or investors
- Testers
- Business analysts
- Technical support
- Legal experts
- System designers
- Documentation experts
- Marketing experts
- Competitors
- Technology experts
- Domain experts
- Regulatory bodies in the industry
- Representatives of trade associations

Table 3.2 Likely concerns and motivations for a range of different stakeholders

Stakeholder	*Possible motivations*
Clients or sponsors	How will I make money from this project? What risks could prevent me getting a return on my investment?
Customers or users (57 varieties)	How will this help me do my job better? How will it help me save time or make money? How will it help me have fun?
Shareholders or investors	Will this initiative increase or decrease the value of my shareholding in the company?
Testers	How do I go about testing this site?
Business analysts	How can I make sure the developers include the results of my business modelling in their design?
Technical support	What are the main problems that users are likely to have with this site? When a user phones me with a problem, how can I tell which page of the site the user is on?
Legal experts	Does the site contain any graphics or information that is copyrighted by someone else? How can I protect the information on the site from being used by other sites?
System designers	How do I go about coding the design? Can I re-use code from other projects? Can I develop new skills on this project?
Documentation experts	How will I teach people to use this?
Marketing experts	How will this initiative increase the company's brand value? How can I use it to collect data about customers?
Competitors	How will this initiative affect my market share? What ideas from this site can I use on my own site?
Technology experts	Does the site use the very latest technology and does it do so appropriately? Are there any security concerns?
Domain experts	If I help design the site, am I doing myself out of a job? Will this de-skill me?
Regulatory bodies in the industry	How does this change the competitive environment?
Representatives of trade associations	How does the site compare with others in the industry? Which is the 'best buy'? Does the site do things in a novel or exciting way?

point since the specific names you have identified will have different and additional concerns.

Prioritise the list

During the design process, you will interact with a number of the stakeholders. Many will give you their opinion of what needs to be in the site, what is wrong with the site and what needs to be changed. All of the feedback that you receive is useful but you may not have time to incorporate all of the suggestions, and anyway some of the suggestions may contradict each other. How do you weight these concerns?

By prioritising the stakeholders, you can decide which stakeholders to pay attention to and which ones are less important to the success of the project.

A simple method for prioritising the list of stakeholders is to use the following technique. First, note the size of the stakeholder group as a proportion of all the stakeholders and define this as Low (L), Medium (M) or High (H). For example, a single person such as the CEO will form a low proportion of the stakeholder group, whereas the end-users will probably form a high proportion. Next, note the importance of this stakeholder group to the success of the project. Once again, define this as Low (L), Medium (M) or High (H). So for example, the importance of the CEO will probably be high, since he or she can cancel the project.

Next, use the table in Figure 3.1 to assign a priority to the stakeholder group. In our example, the CEO will have a priority of 3 as highlighted in the table.

Continue this exercise for all of the stakeholders. This will provide you with a prioritised list to help you decide whom to focus on and why.

Devise a management strategy

The final step in this stakeholder analysis is to consider how you will manage each of the stakeholders' concerns. For example, any investors in your e-commerce development will be concerned about getting a return on their investment.

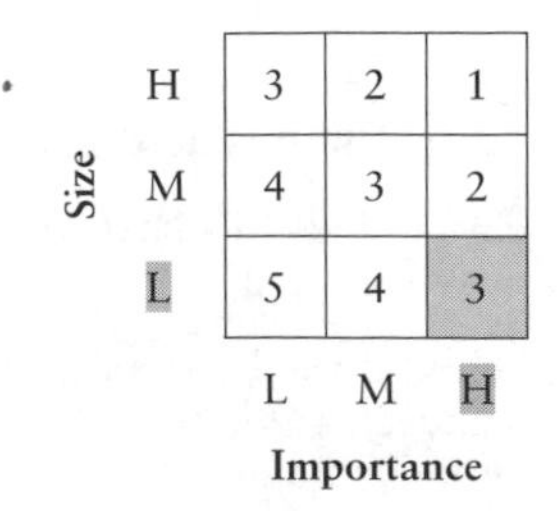

Size \ Importance	L	M	H
H	3	2	1
M	4	3	2
L	5	4	3

Figure 3.1 Example of assigning priority – 'CEO'

Consider as an example a fictitious project termed 'TicketTeller'. TicketTeller is a web-based theatre ticketing system. Assume that the inspiration for a web site emerged from the observation that foreign tourists are reluctant to buy tickets from theatre box offices because of language difficulties. So the aim of the initiative is to provide a web site where customers can read full London theatre listings and buy tickets for theatre performances. The main access point will be through a web browser and kiosk access may be a later development (kiosks could be placed in London theatres and tourist attractions).

Imagine that one of your investors looks at some early design ideas and asks for changes to the user interface that you feel are likely to hinder usability. As a specific example, consider a request to make a theatre logo twice as large. Simply pointing out the possible usability problems this may cause may have little impact (unless this investor especially values usability). To manage the investor properly our concerns need to appeal to the investor's motivations.

For example, we might point out that the cost of increasing the size of the logo is that this will move the 'theatre listings' functionality lower on the screen. To see this functionality, customers will then have to scroll. Many customers will scroll, but a percentage of them may not. If 10 per cent of customers do not scroll, what might the costs of this be?

If it fits the investor's motivational style, we can be very specific at this point. Assume the site plans to have, on average, 10,000 unique visitors per week. Assume further that, of the ten per cent of these (that is, 1,000 visitors) who fail to see the listings, a small number – say 20 per cent or 200 visitors – quit the site assuming that the site does not provide this functionality. (Presumably the other 80 per cent do not need theatre listings or they will stumble around the site and find them.) Assume further that five per cent of the 200 visitors who quit (ten visitors) would have made a purchase at an average cost of £100. The cost of this problem is therefore 10 × £100 or £1,000 per week, which scales to £52,000 per year.

We have now addressed this stakeholder's concerns with a management strategy focused on describing the return on investment.

How to present the information

The approach advocated in this book is to use a form-based approach. This will ensure that you address all of the main issues. Of course, the form itself is not the critical part of the process; so long as you make an explicit note of your answers, any format will do. However, a form-based example is shown below (and a template is provided in Appendix 1). The example in Table 3.3 shows one entry in the table for the 'TicketTeller' system.

Table 3.3 Example of stakeholder analysis of one stakeholder

Stakeholder Roles and Main Concerns	Management Strategy
Shareholders/Investors The main investor in the project is a consortium of London theatres. The lead investor is J. Williams whose main concern is getting a return on investment.	Use cost-benefit analysis to illustrate the financial impact of design decisions. For example, assuming 10,000 visitors per week and an average transaction of £100, an increase in conversion rate from 5% to 6% will increase turnover by £10,000 per week or £520,000 per year.

Summary

- All web projects have a number of stakeholders who can help the project succeed – or fail.
- You should create a list of stakeholders, prioritise the list and devise a strategy to manage each stakeholder group.

Further reading and weblinks

Books and articles

Neely, A., Adams, C. and Kennerly, M. (2002) *Performance Prism: The Scorecard for Measuring and Managing Stakeholder Relationships*. Upper Saddle River, NJ: Financial Times/Prentice Hall.

Web pages

Overseas Development Administration (1995) 'Guidance note on how to do stakeholder analysis of aid projects and programmes'. http://www.euforic.org/gb/stake1.htm. Not a web project, but an excellent case study containing checklists and worked examples of a stakeholder analysis.

4 Write the site mandate

In every e-commerce development, there comes a time when a designer wants to add a 'unique' feature. The precise feature differs from project to project but these features share some common characteristics: it takes an experienced programmer or designer to implement; the feature does not appear on many other web sites; and the feature takes time for customers to learn how to use it. The programmer or designer has usually tried out the feature on a colleague at the next desk who has been suitably enthusiastic. There is a murmur going around the office: 'Have you seen Jeff's new menu? It's cool!'

Perhaps you have had these moments on your own project. I always find myself cast in the role of bad guy when this situation comes up. While everyone else is saying 'Wow!', I am wondering if the typical customer for this site can be bothered to download the plug-in, configure the browser properly and then spend time learning how to use this flashing 3-D animated menu. I say something like, 'Hey it's very impressive, but you need to remember that our customers aren't high-energy physicists you know'.

Which is ironic, because back in the very first days of the web, web users *were* high-energy physicists. Researchers at CERN, the European High-Energy Particle Physics lab in Geneva, needed a collaboration tool. Tim Berners-Lee gave them the World Wide Web. The people that designed the sites were high-energy physicists and the people who used the sites were high-energy physicists. Both the designers and the users were hewn from the same social and intellectual block: designers and users sat at similar computers, shared the same goals, and wanted to solve similar tasks. If a physicist was working on a page and wanted to check that a fellow physicist would understand it, he just needed to turn to the next desk and ask whoever sat there. This made good web design very easy.

In the early days of the web, 'next desk design' worked very well. But in today's web environment, 'next desk design' is lethal. Your customers may very well have different goals to your colleague at the next desk and they are probably a different kind of person.

This makes good web design very difficult. You may have a feature, or even an idea for a site, that you think is great. You try it out on people at the next desk, and they think it's great too. But the idea flops because the customers of the site (14-year-old skateboarders, general practitioners, parents with young kids or whoever) cannot or will not use it. These people are not at the next desk – they may never even sit at a desk. But it is these people who will decide if your site will win or lose on the web.

In the early days of web development, not having a clear idea about your customers or the goals of the site were acceptable – because just having a web site was an achievement. 'Brochureware' sites (ones that simply reproduced the company brochure) were endemic. But in today's web, customers are more savvy and have much higher expectations from sites. So it is important to identify *why* the site is being developed and state precisely *what* the site aims to achieve.

You may be the project manager. If so, you already have a good idea of why the site is being developed. So you can skip this step, right?

Wrong. The purpose of this step is to make sure that *everyone* on the project team shares your vision for this project. It is not unusual to find web development projects where the CEO wants a site that allows people to order on-line, but the marketing team thinks that the aim of the site is to collect customer relationship management (CRM) data on customers. Moreover, developers think the aim is to get people to register.

So the site mandate needs to be stated explicitly – in writing. On the other hand, we want to avoid bureaucracy and rigid processes. One effective way to achieve both goals is to use a form-based template that allows the main points to be quickly captured and circulated without drowning anyone in project documentation. This process has a number of benefits:

- nobody misses important steps;
- communication amongst the project team is strengthened;
- quality is improved;
- an audit trail is in place (critical to achieve compliance with ISO 13407).

None of the processes described in this book will limit designers' or developers' creativity and they have been tailored to help teams work quickly. Let us now look at the kind of information we need to specify in the site mandate.

Project or site name

Every project or new development needs a name. This does not need to be the final name for the site, just something that provides a rallying point for the project team.

Characteristics of the site

We now need to describe the main characteristics of the site. What kind of site is it? Does a similar site already exist? If so, what is currently good and bad about it?

This may be a new development. If so, what evidence do you have that customers will be interested in it? Are there any particular technological constraints that need to be acknowledged? For example, your system may be aimed at WAP phones in particular, or handheld PDAs in general, or perhaps you anticipate customers using your site through a regular web browser. Make your assumptions explicit now before you find the development team wasting effort researching a technology platform that was never envisaged.

This is also the point to make a first guess about the number of unique visitors and the conversion rate you will aim for. The conversion rate – or 'lookers to bookers' ratio – is the number of visitors to your site who do what you want them to: for example, buy a product, register for special offers or sign up for your mailing list. Conversion rates in the web industry as a whole tend to be in the low single percentages (around two to five per cent), although some Internet-only retailers such as Amazon.com are reported to have conversion rates in excess of ten per cent. Conversion rate is discussed in more detail in Chapter 13.

Business model

This is the time to make explicit how the site will make money. For example, the site may try to support itself through subscriptions (if so, how much?), by accepting advertisements (if so, from whom?), by selling a product or by charging commission. Alternatively, the site may not be designed to make a profit, but perhaps to generate qualified leads for the business. In this latter case, the business model is that funding for the site will come from within the company (perhaps earmarked as a sales and marketing expense).

Business and brand objectives

Before considering the objectives of the site, we need to first acknowledge the objectives of the organisation. What market is this organisation in? What are the organisation's medium- and long-term aims? Why does the organisation feel that the web offers a route to achieving these aims?

One way of answering these questions is by exploring the company's brand values. Brand value is quite different from brand awareness: brand awareness merely shows that customers know you exist. That is no guarantee that they will do business with you. Measurement of brand value is an inexact science, but it is generally accepted that strong brands increase sales and earnings. Brand, as we have said before, means more than a good logo. Customers' perceptions of a brand are constructed from their entire experience with an organisation. This means the shop (if the organisation has one), the employees, the customer service representatives, the charisma and character of the CEO, and of course the web site: how easy the site is to navigate, how simple it is to contact someone, how straightforward it is to return a faulty product.

Appropriate site design can reinforce these objectives. Inappropriate design can contradict them. So if your brand value is 'easy to do business with', you better make sure that customers can easily place an order.

Key site objectives

What are the key success factors for the web site? Is it important that the web site makes money? If so, where from? What are the short-term and long-term goals of the site?

What we need here is an aspiration statement to answer the question: 'what needs to happen for the site to be considered a success?'. The site's objectives should fit with your company's business objectives, so you will need to identify your company's business objectives and the brand values that the web site needs to communicate.

This high-level objective should:

- reflect the customers' total experience with the site;
- convey the value that the site will provide for both the customer and the organisation.

Typical themes for such objectives might include:

- increase e-commerce revenue;
- improve lead generation;
- generate new customers;
- extend reach into new markets;
- improve shareholder value;
- collect CRM data on customers;
- reduce costs;
- forge relationships with other businesses;
- build in-house expertise;
- retain staff;
- improve brand positioning with awards or press coverage;
- increase market share;
- improve customer service;
- gain competitor advantage.

Planned target market

The next step is to specify the customers themselves. At this point, a general description will do, capturing the market sector if it is targeted at a narrowly defined customer group (for example, 'architecture'). Also describe the market size and the geographical distribution of the market. If the site is aimed at a mass-market, make it explicit. In the next two chapters we will describe the market, and the target customers, in a lot more depth.

Value proposition

Just as we need to provide a short description of the customers for the site, we need to describe the value that we expect customers to get from using this site. If there are competitor web sites, what will distinguish this site from the competitors? How will this site make customers' lives easier? Why will people visit the first time and why will they come back?

Likely functions

Here we describe (in broad outline) the probable functions the web site needs to have to deliver those benefits. Try to describe these in terms of what customers want to do (functions), not in terms of an implementation (features). For example, 'the customer should be able to navigate to the main areas of the site from any page' is a statement of function. A left navigation bar, frames, a pull-down menu, or some other implementation is a feature.

Competitor web sites

Competitor sites provide a rich source of information at this early stage in the development of the web site. They will give you an idea of the market itself – what people are willing to buy – as well as the way people may like to have this information delivered. Identify the market leader and the main functionality supported at that site.

You should be aware of some risks in this approach too. Your competitors may have developed their web site by copying the ideas of others and as a consequence the design area has become stale. This could be the opportunity to develop a completely different design.

Finally, what if there are no competitor sites? In this case, look to see if there is a similar, manual task that the site is automating. For example, if you are designing an on-line personal organiser, you could look to see how people manage their time with other electronic organisers.

How to present the information

Appendix 2 provides a template for the site mandate, and Table 4.1 shows a worked example for the 'TicketTeller' web site introduced in the last chapter.

Summary

- The purpose of the site mandate is to make the purpose, aims and objectives of the site explicit.
- The site mandate briefly describes: the project or site name; the characteristics of the web site; the business and brand objectives; the key site objectives; the target market planned; the value proposition; the likely functions; and any competitor web sites.

Table 4.1 Site mandate for the 'TicketTeller' web site

Characteristics	Description
Project or site name	TicketTeller
Characteristics of system	Web-based theatre ticketing system. The system is intended to support the standard method of ticket sales (mainly by telephone). The current system is very flexible but our research shows that foreign tourists are reluctant to use it because of language difficulties. The main access point will be through a web browser (i.e. no WAP access is expected); kiosk access may be a later development (kiosks could be placed in London theatres and tourist attractions). After release, we expect 10,000 visitors per week and aim for a conversion rate of 5 per cent.
Business model	The site will make money by charging customers a booking charge (either a flat fee or a percentage of the transaction).
Business and brand objectives	Increase theatre occupancy and increase London theatres' 'Wallet share' of tourists' money. English theatres have a distinctly 'cultural' brand and this should be reflected in the design of the site.
Key system objectives	The majority of theatregoing tourists should prefer to use this system rather than the telephone. The number of foreign tourists at London theatres should show a significant increase after the first year of operation. Good communication is a key aspect of theatre and this should be reflected in an easy to use site.
Target market planned	Mainly aimed at tourists to the London area (some of whom may have language difficulties). The majority of these are Japanese. A second important market is the frequent UK theatregoer.
Value proposition	No queuing. No language difficulties.
Likely functions	Full London theatre listings and the ability to purchase all major ticket types (excluding those needing discount cards). The site should accept payment by credit card and provide access in multiple languages (especially Japanese).
Competitor systems	There are no direct competitors. Several theatre-listing sites exist but these do not allow bookings to be made. We will check other ticketing systems (e.g. airline sites) for design ideas and management of multiple languages. The theatre ticket-buying process can also be analysed by considering the current telephone booking system.

Further reading and weblinks

Books and articles

Burdman, J.R. (1999) *Collaborative Web Development: Strategies and Best Practices for Web Teams*. Boston, MA: Addison-Wesley.

Chase, L. (1998) *Essential Business Tactics for the Net*. New York, NY: John Wiley & Sons.

Web pages

AWARE (undated) 'Competitor analysis – a brief guide'. http://www.marketing-intelligence.co.uk/aware/competitor-analysis.htm. 'Collecting Information on competitors can be likened to prospecting for gold. Nuggets are a rarity. The prospector will need to sift through a lot of soil, to find the few grains of gold which make the task worthwhile. Occasionally, the prospector will even be tricked by iron pyrites – or "fool's gold"!'

Robert Manning (2000) 'Internet branding and the user experience'. http://www.clickz.com/brand/branding/article.php/822571. 'In the Internet space, branding means creating a great user experience. Internet branding moves beyond logo, tagline, key messages and graphic identity into the customer's real-time interaction with the brand, for the entirety of the online experience.'

Russell, G. (2001) 'When requirements interfere with usability goals'. http://www.ganemanrussell.com/newsletter/07012001.html. 'Designing a usable web site requires more than being able to define the business and project requirements. These goals must be in synch with usability principles and goals, or the design process may suffer.'

Sean Carton (2000) 'Brand is back'. http://www.clickz.com/tech/lead_edge/article.php/821891. 'While many consumers give lip-service to price when describing what drives them to a particular site, it's the unique features that make shopping easier and more satisfying – such as customer support, value, and overall satisfaction – that keep them coming back.'

UsabilityNet (2001) 'Competitor analysis'. http://www.usabilitynet.org/methods/planningfeasibility/competitoranalysis.asp. A 'how-to-do-it' tutorial.

5 Segment the market

Market segmentation is a process by which a market is divided into standardised groups or 'segments'. The goal is to be able to produce a different 'marketing mix' that will appeal to each segment – such as different promotions, different communication channels and even different site designs.

The aim of this chapter is to describe how market segmentation will help you focus your web site on specified groups of customers. Too often sites are designed to appeal to everyone, with something for everybody: a registration page built around the wants of marketing, a thesaurus-based search engine that appeals to developers, a Flash animation that a graphic designer thinks is 'way cool'. The problem is that a site with something for everybody has got everything for nobody. In contrast, market segmentation will help you decide who the site is aimed at and how the site should be delivered; and equally importantly, it will also help you decide who the site is *not* aimed at.

What are the benefits of segmentation?

Market segmentation provides a number of key benefits. First, it will help you avoid 'feature creep': less functionality is needed to meet the needs of a specific customer group than to meet the needs of everyone. Second, it helps constrain the visual design: segmentation will help avoid sending the wrong message. Third, it helps you discover new markets, or markets that are poorly served at the current time. Finally, it provides efficiency savings by identifying the best segments for your site. This means marketing money can be spent attracting the right customer group.

But perhaps the biggest advantage of segmentation is that you are able to give customers what they want. Imagine for a moment that we are in the business of providing summer holidays. Three separate customers come into our shop: Karen, looking for a clubbing holiday with her friends; Ian, looking for a beach holiday with his family; and Harry, looking for a cultural holiday in Europe with his elderly wife. We could try to satisfy their needs with a single holiday destination, but the result of trying to satisfy all of them would result in satisfying none of them. Web sites are no different, and by segmenting

Box 5.1 Segmentation and accessibility

I have sometimes come across examples of people using market segmentation as an excuse to exclude users with disabilities from their site. Since the market segmentation mantra is 'design for some', how can this be reconciled with the notion of 'design for all'?

The main reason why sites tend to be inaccessible to partially-sighted customers is because the visual design excludes them. But if the site is professionally designed then the visual design of the site should be entirely independent of the content. For example, poorly designed sites may render text in an image, which means that the text cannot be enlarged by a conventional browser and makes the text unreadable if the customer is using a text-only browser. There are relatively simple fixes to these kinds of problem, such as using the HTML <ALT> tag in this specific example, and cascading style sheets in general offer designers an ideal way to separate visual design from the underlying content. Designers that produce inaccessible pages cannot hide behind the notion of market segmentation.

There are additional business reasons to drive accessible sites. Improved accessibility benefits all customers: in the real world, consider how pavement cut-outs for wheelchair users help parents with buggies or people wheeling a suitcase. In the virtual world, some users may access your site with a text-only device or with a small screen (such as a PDA): these users are in exactly the same position as a partially sighted customer using a text-only browser and they will value an accessible site.

Moreover, people with disabilities cut across many market segments because disabilities are so diverse. Over 8.5 million people in the UK are registered with some form of disability (this includes over 2 million people with a visual impairment). In particular:

- 8 million people suffer some form of hearing loss.
- 1 million people have learning difficulties.
- Over 7 million have literacy problems.

Impairments can take a variety of forms and may exist together in combinations.

the market you will be able to decide which specific groups of customers you intend to aim your site towards.

How do you segment a market?

If your web site targets an existing customer (or a competitor's), information that describes these customers may already be available. So start by using the profiles of market segments that already exist within your organisation.

When you are thinking about end-users (i.e. the people who interact with your web site to get work done), these categories may be useful:

- novices – curious, but afraid to make mistakes and reluctant to ask for help;
- occasional customers – can do particular tasks well but reluctant to explore the whole web site;
- transfer customers – already know how to use a similar web site;
- experts – have a thorough grounding of product concepts, hate to waste time and want shortcuts to make their work more effective;
- rote users – simply follow instructions and do not understand the concepts underlying the web site.

Segmentation using the 'chasm' model

I now want to describe a unique approach to market segmentation designed specifically for the high-technology sector (of which web sites are a part). In the 1990s, Geoffrey Moore wrote an influential book titled *Crossing the Chasm*. In this book, Moore describes his consultancy work with a number of high-tech companies. He argues persuasively that many companies fail in the very early stages of development because they do not adopt a strategy appropriate for the phase of technology adoption in which they find themselves.

The technology adoption lifecycle is usually characterised as a bell-shaped curve, illustrating the fact that new technology is quickly adopted by some (the innovators and early adopters) but more slowly by the majority of customers. Finally, all products (unless they are re-invented) will enter a phase of decline as new entrants push them out. Moore's Technology Adoption Life Cycle is shown in Figure 5.1.

The first phase of the lifecycle is the 'early market'. In this phase, the product is new and exciting, but is rarely a targeted solution to a particular problem. More often, it is a technology in search of a solution. Moore argues that it is at this point that many new products fail: they reach the 'chasm' and are not adopted into the mainstream. Products that manage to successfully negotiate this phase do so by customising their product for a niche group of customers: by adapting the product to become a 100 per cent solution to the niche customer's needs. Moore describes this phase as the 'bowling alley' since customer niches need to be identified and met. The 'tornado' – a period of mass-market adoption – then sweeps up successful products, until they reside on 'Main Street' where the product is standardised and the aim is to turn the product into a commodity. Finally, the product reaches 'end of life', when it is replaced by new technologies.

For our purposes, Moore's key contribution is to point out that the different phases of this lifecycle require different – indeed, sometimes contradictory – strategies:

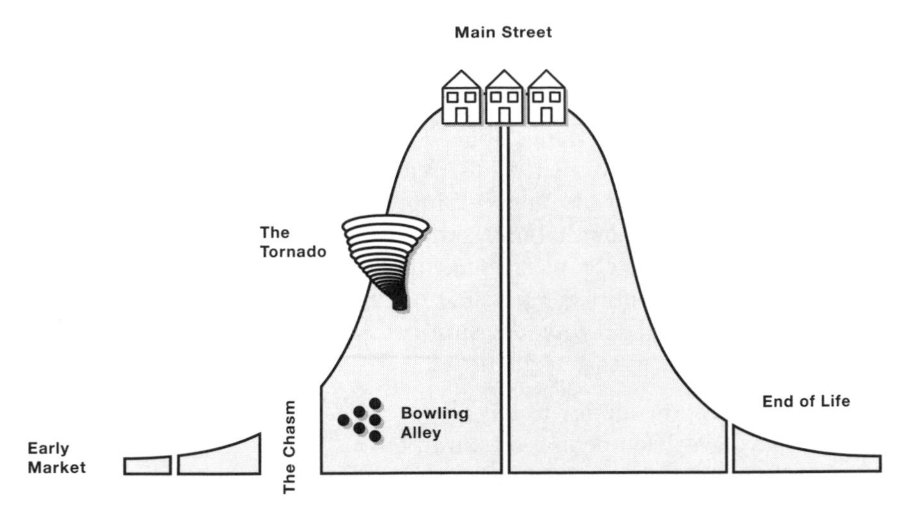

Figure 5.1 The technology adoption lifecycle (after Moore)

> As markets move from stage to stage in the Life Cycle, the winning strategy does not just change – it actually *reverses* the prior strategy. The very skills that you've just perfected become your biggest liabilities; and if you can't put them aside to acquire new ones, then you're in for tough times.
>
> G. A. Moore, 1995: jacket

In particular, Moore points out that the profiles of customers are quite different in each of the different lifecycle stages. In the 'early market', the first customers are technology enthusiasts: these customers love technology for its own sake. Although they rarely have the money to invest in the product, they can influence the second breed of customer within the 'early market': the visionaries. Visionaries have money to invest in the product and want to use it to achieve competitive advantage. However, visionaries demand special modifications that are of little interest to the majority of customers.

Getting the ear of technology enthusiasts and visionaries is not that difficult. These people consume technology and there will always be someone, somewhere, who finds your new invention interesting. The key to success is to cross the 'chasm'. Moore describes this concept in the following way:

> The idea of the chasm is a simple one. It says that whenever truly innovative high-tech products are first brought to market, they will initially enjoy a warm welcome in an early market made up of technology enthusiasts and visionaries but then will fall into a chasm, during which sales will falter and often plummet.
>
> G. A. Moore, 1995: 19–20

To cross the 'chasm', your product must be of interest to the next group of customers: pragmatists.

If technology enthusiasts want to use technology for its own sake and visionaries want to use technology to get things done, then pragmatists – well, they just want to get things done. This group of customers are looking for a complete solution to their needs. Within the 'bowling alley' this means identifying a key group of customers – Moore terms this a 'beachhead segment' – and delivering a solution that meets their need in its entirety. This group of customers becomes the complete focus of the effort to cross the Chasm.

Once you have identified a possible beachhead segment, Moore recommends that you validate your decision by asking the following questions about the customers in that segment:

- Do they have the money to buy?
- Do they have a compelling reason to buy?
- Can we deliver a whole product that will fulfil their reason to buy?
- Is there no entrenched competition to prevent us having a fair shot?
- If we win this segment, can we leverage it to enter additional segments?

Once that segment has been won, you can move onto the next bowling pin. You do this, not by re-designing your web site from scratch for a new segment, but by offering the same functionality to slightly different segments or by offering slightly different functionality to the same segment. For example, Amazon.com began life as a web-based bookseller (functionality), with a product aimed mainly at web-savvy customers (the beachhead segment). The company then diversified in two ways: first, it offered the same functionality to slightly different segments: it sold books to anyone who could use a web browser. Then it offered slightly different functionality to the same segment: for example, Amazon.com began to stock software and CDs.

Once on 'Main Street', the goal is to sell to conservatives. This group of customers are reluctant to use technology and do so only under some kind of pressure. This group of customers are price sensitive and very demanding. They have no interest in the underlying technology and just want something that works – a commodity.

The final group of customers choose to buy when the product is so ingrained that it is in decline. These are the sceptics, and have so little influence on product acceptance that they can be ignored.

Table 5.1 summarises this analysis of Moore's work as it applies to the customer-centred design (CCD) process.

Summary

- Market segmentation is a process by which a market is divided into standardised groups or 'segments'.

Table 5.1 Customer-centred design and the technology adoption lifecycle

	'Early Market'	'Bowling Alley'	'Tornado'	'Main Street'
Customers want	To figure out how the site works To experience new technology To use basic capability but with unique customisation features	A total, niche-specific solution to a critical problem	To get their work done better and faster Stability and consistency	Basic product at cheapest price Niche-specific extensions
Customer-centred design means	The site works	Profound understanding of the customer's needs Task-oriented design	Quick to learn Fast Powerful	The product is invisible
Site should focus on	Eradicating major bugs Customising existing features	Adding the right features in the right form	Performance Fixing bugs Lowering support costs Scalable (to handle increased load)	Cost of ownership New markets ('whole product +1')

- Market segmentation provides a number of key benefits, such as avoiding feature creep and providing pointers for the visual design of the site.
- The 'chasm' model separates the technology adoption lifecycle into a number of phases, each of which requires a different strategic approach.
- Focusing on customers in a beachhead segment and delivering a solution that meets their need in its entirety is a key strategy for a successful web site.

Further reading and weblinks

Books and articles

Frigstad, D. (1994) *Know Your Market: How to do Low-Cost Market Research*. Central Point, OR: Oasis Press.

Parker, R. (1997) *Roger C. Parker's Guide to Web Content and Design: Eight Steps to Web Site Success*. New York: MIS.

Web pages

Cyberatlas (2002) http://cyberatlas.internet.com. Market research and Internet statistics focussing on e-commerce.

Usability.gov 'Statistics & market research'. http://usability.gov/statistics/index.html. 'Links to statistics on user behaviours and trends, search engine use, and user and Internet survey findings.'

W3C (1999) 'Web content accessibility guidelines 1.0'. http://www.w3.org/TR/WCAG10/. These guidelines explain how to make Web content accessible to people with disabilities.

Part II
Step 2
Build the context of use

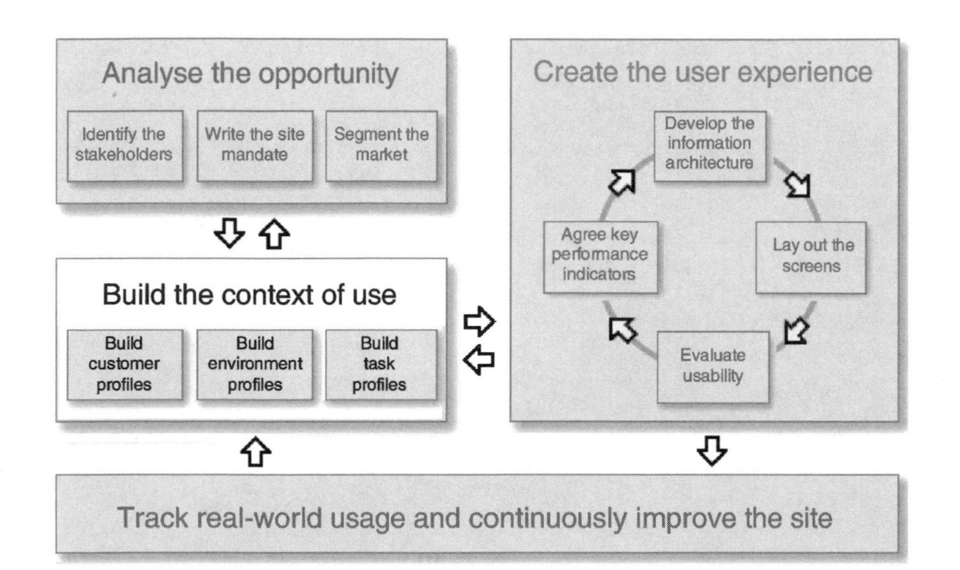

6 Build customer profiles

Successful businesses understand their customers. This is almost a cliché, and I doubt that anyone would disagree with it. Indeed, everyone involved in the design of a web site will tell you that they know all about their customers. As an exercise, I once compiled a composite description of the customer of an interactive TV service based on the comments of the design team. The customer was between the ages of 12 and 80, male or female, with a range of computer and web expertise. This is hardly the type of specific information that you need to specify a beachhead segment, and a major part of this assignment was helping the client appreciate that 'something for everybody' translates to 'everything for nobody'.

The point is that although we all agree that it is important to understand our customers, few of us actually take the time to collect the information that we need to understand them. Interestingly, the same people would never take the same approach with new technology. Nobody would say, 'Oh yes, the cyberfrangler plug-in, let's have one of those'. In fact, the company would probably recruit an expert in cyberfrangler authoring who understands the technology inside and out before taking any risk that the new technology might fail.

Perhaps the reason why few people take the time to collect detailed information about customers is because they genuinely feel they understand who the customers are. A skeleton profile of an imaginary customer can be built very quickly, based on false assumptions and anecdote. And this kind of back-of-a-cigarette-packet analysis gives you more time to read up on the cyberfrangler plug-in. In contrast, collecting real data about real customers takes time. It also means you have to speak with customers and this can be a very uncomfortable experience if they are familiar with your web site and don't like it.

Another problem is that marketing departments define the customer as the person who makes the purchase. This definition of the customer is appropriate for marketing but it does not help us develop an easy-to-use web site. This is because the person who makes the purchase may never actually use the site. For example, I worked on a web-based health benefits portal

some time ago. The site was aimed at fairly large, blue-chip companies. When a company subscribed to the site, it accessed the site from within its intranet. The purchasers of the site were usually managing directors, or at least the board director with health and safety responsibility. But employees used the web site. Having lots of information about the managing director (marketing's customer) told us nothing about how the site should be designed for end users.

Building pictures of customers

The aim of this chapter is to provide you with techniques you can use to find out more about your customers. You will learn techniques for building pictures of your customers and the environments in which they work. You will gain an understanding of customer requirements in the context of their job. And you will learn what it is that customers actually want to do with the site.

We will achieve this by reviewing something we call the 'context of use'. This term was coined in a usability standard published by the International Organisation for Standards (ISO 9241-11). The term helps capture the idea that usability is context dependent. It makes the point that usability is not a feature of products or web sites. Rather, usability is a property that emerges when we design a product that sits in the triangle made by customers, their tasks and the environment in which they use it.

Customer profiling

There are many ways to get a better understanding of your customers. I have divided them into two sections: those that require 'direct' contact with customers and those that do not.

Indirect methods

Use indirect methods to understand your customers when it is difficult to meet with them face-to-face.

Surveys

Surveys are widely used on the web and elsewhere. They are a fairly easy way to collect lots of data at very low cost. This particularly applies to surveys that are administered via a web server, since the computer can begin to carry out some rudimentary analysis. You simply sit back and watch the bar charts change in front of your very eyes.

But they also suffer from disadvantages. First, the questions need to be determined in advance. To make sure they can be completed quickly, surveys tend to be populated by check-box type answers. This means respondents will fail to tell you something germane if you don't ask them the question.

My favourite example is the failed business that carried out an extensive survey of potential customers before launching the product. The survey results were very positive but the site failed. It used a subscription-based business model, but the survey designers forgot to ask if people would actually pay for the service. Respondents had assumed that, this being the web, it was probably free anyway. This means that you can be badly misled by survey results if you fail to ask the right question.

A second problem is that return rates are usually low. Even with incentives (such as discount vouchers for completing the survey), many people are reluctant to take time out to complete surveys, even very short ones. You therefore run the risk that your completed surveys come from 'professional survey takers' and are not representative of your customers as a whole.

A third problem is that it is sometimes difficult to follow-up respondents. Unlike in an interview, when you can use a follow up question to better understand the issue, with a survey you get just one chance.

The main reason for the popularity of surveys stems from the fact that they are so easy to design and administer. However, as hordes of survey designers will tell you, it is not easy to design a *valid* survey. Quickly cobbled-together surveys are invariably of dubious merit and can cause significant bias.

An example of bias in practice can be seen in this example from MORI (*British Public Opinion*, 21(10) Dec 1998). They posed the following question:

> As you may know, there has been a proposal that ITV should change its line-up of programmes in the evening. One part of this would be to move the main news from 10 p.m. In general, how strongly would you support or oppose moving the news from ten o'clock to allow ITV to show other programmes?

The results showed that 25 per cent supported the move and 41 per cent opposed the move. The analysts then tried a different question:

> As you may know, there has been a proposal that ITV should change its line-up of programmes in the evening. One part of this would be to move the main news from 10 p.m. If there were no evening news at ten o'clock, ITV could show a wider range of programmes between 9 p.m. and 11 p.m., including drama, films, documentaries, current affairs, sport and comedy. In addition, two-hour feature films not suitable to be shown earlier in the evening could be shown without a break for the news. In general, how strongly would you support or oppose moving the news from ten o'clock to allow ITV to show other programmes?

Now the results showed a dramatic increase in support for the move (43 per cent) and a substantial decline in opposition (30 per cent).

By carrying out methodical work, MORI were able to identify sources of bias in their questions. But this is rarely the case with surveys that are quickly put together.

The learning point here is *not* that surveys do not have a role to play in helping you understand your customers. What this teaches us is that you must be very careful about the questions you ask and the way you phrase them. First, try not to ask questions that are a matter of opinion. Instead, use surveys to get an understanding of your customers' behaviour. So for example rather than ask the question: 'Would you use a web site that lets you manage your diary on-line?' (an opinion question) ask 'Do you use an electronic diary?' (a factual question). These types of survey are especially useful for aggregating facts about your customers' behaviour and lifestyle.

Web site logs

Every time one of your visitors views a page or loads an image from your site, your web server makes a note of this in a logfile. An excerpt from a web site log is shown below:

```
202.9.141.102 - - [28/Oct/2000:05:41:11 +0000] "GET /articles/callcentre.html HTTP/1.1" 200 9226
```

The logfile comprises thousands of lines, but just one is shown above. The line above is broken into seven fields (see Figure 6.1). Table 6.1 explains the content of each of these fields.

At first glance, this looks like a potential way to find out what customers are doing at your site. And indeed, logfiles are a great way to develop 'Boss factoids'. ('Boss factoids' are easily digestible figures that you can use to describe your web site's performance to your boss.) Examples include:

- the 'top ten' most popular pages;
- the number of hits;
- the busiest days and times;
- the number of requests by domain (e.g. .com .uk, etc.);
- average number of page views per day.

However, there are some pitfalls in interpreting these data. Take the IP address for example. This may not correspond to the same person, or even a person at all. An IP address could represent a 'spider' (an automated browser) or a proxy server (which in the case of an internet service provider could

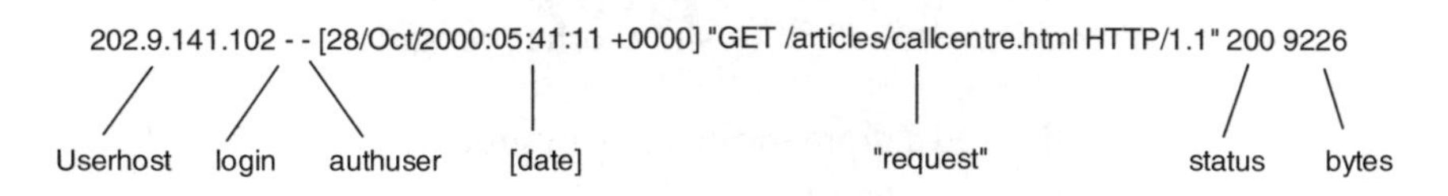

Figure 6.1 How a logfile is constructed

Table 6.1 How to deconstruct a logfile

This part of the logfile ...	*... refers to*
202.9.141.102	userhost: the machine name or IP address of the computer that made the request
-	login: the user's remote login, if provided. The example does not provide that information, so all you see is a dash to show that there is no data for that field.
-	authuser: the user's log in name, if the site is authenticated. Our example shows an open site, so there is no authuser data.
[28/Mar/2002:05:41:11 +0000]	[date]: the date and time the request was made
"GET /articles/callcentre.html http/1.1"	"request": the path to the information resource requested by the client. This field also includes the protocol requested by the browser. In the example above, the browser requested the callcentre.html page in the articles directory, using http 1.1.
200	status: the http status code of the request (200 in our example).
9226	bytes: the number of bytes in the document that was requested

represent a large number of different customers). So these data do not provide an infallible understanding of the people visiting the site.

To use these data to understand your customers, you will need to read between the lines: logfile analysis is as much of an art as a science. For example, assume that you want to find out the type of network connection that your customers use (low-bandwidth dial-up connection or a high-bandwidth T1 line). This information cannot be explicitly logged, but you can get some idea by looking at the busiest days and times. This analysis will be based on *the balance of probabilities* and you need to make a number of assumptions. Here is an example of how you can do this analysis.

First, you need to divide customers up into time zones. The time given in your logfile is the time on the server, and this may be many hours ahead or behind of the customer. You can make a rough stab at this by mapping the last part of the IP address to particular countries (this is not a fail-safe method – not all .com domains are based in the US – but remember, this is an art, not a science). If the IP address is numeric you could carry out a 'reverse DNS look-up' on the Internet to find the domain, but this will take a long time if you do it by hand so for efficiency it is probably best to exclude numeric IP addresses from your analysis.

Once you have converted the server times in your logfile to the customers' local times, you can take the next step. Customers visiting your site between 6pm and 8am local time are probably accessing it from home. Between 8am and 6pm local time they are probably accessing it from work. Let's assume that for our logfile, 80 per cent are accessing the site from home and 20 per cent from work.

Finally, we assume that most 'home' customers will probably be using low-bandwidth connections, and most 'work' customers will probably be using high-bandwidth connections.

This piece of interpolated data, although subject to some degree of error, is now much more valuable than the raw data in the logfile because you now understand something specific about your customers. For example, if around 80 per cent of your visitors are using dial-up connections they will not want to hang around at your site since they are probably paying by the minute for the connection. They will also not want graphically-rich pages because these will take an age to download.

A second approach is to analyse the data in the form of 'session logs'. With a session log, you collect together all of the page requests made from a particular IP address. Assuming that these all represent the same customer, you can then stitch the requests together and see the path taken through your site by that customer. This is a useful way to discover pages where customers leave before completing their tasks or to identify 'dead-end' pages (the last page that a customer views before leaving the site).

Artefact analysis

With artefact analysis, you collect objects related to customers' tasks. The term comes from archaeology where experts uncover objects during a dig and then have to piece together the objects and try to understand for which tasks people or communities used the objects.

A typical shopping artefact with which we are all familiar is a shopping list. Assume you had access to a customer's grocery shopping list: what could you learn from it? First, you could look at the way items on the list are ordered. If the customer groups certain items together (for example, bread, milk, eggs) this tells you something about the way the customer groups items in his or her mind. This information may help you structure information on the site so that these items could be found in the same place. But more radically, there may be ways that you can use the artefact to match the customer's task. For example, rather than make the customer navigate through a hierarchy for each item and add it to the electronic shopping cart, the customer could simply type in the complete list. The webserver could then carry out the laborious task of matching the various items with the items in its inventory using a thesaurus (for example, 'loaf' for 'bread'). Let's say that a customer's shopping list shows that the customer wants to buy bread, milk, eggs and wine from a grocery store. Rather than navigate to and add separate items in the cart, the customer could simply type in these four items and the software could search for matching items. The customer will need to specify brown or white bread, sliced or unsliced etc. but this can be left until later.

Artefact analysis cannot be your only method of data collection, but it can provide invaluable insights into the way customers currently do the task.

Indirect methods: summary

SURVEYS

- Advantages: Good for collecting a high volume of data at low cost; effective tool for measuring attitudes and obtaining factual information (demographics, equipment); good for scoping investigations (e.g. likes, dislikes).
- Disadvantages: Topic must be predetermined; follow up is difficult; return rates are often low.

WEB SITE LOGS

- Advantages: Provides objective data with minimal cost and intrusion; can cover large periods of productive use; good supplement to observation; can help in troubleshooting defects.
- Disadvantages: Difficult to analyse data; requires web site analysis tools; provides no understanding of the people doing the work; only works if there is an existing site.

COLLECTING ARTEFACTS

- Advantages: Collecting artefacts provides physical examples of portions of a customer's task; provides a good way of concretely communicating needs or definitions to other developers; low in cost.
- Disadvantages: Customers may not be willing to provide examples; will only cover limited areas of task; cannot be the only method of data collection.

Direct methods

Use direct methods when you can actually see customers in action.

Observation

Observation techniques are characterised by watching customers go about their tasks. Customers might be using your site or a competitor site or they may be carrying out the activity in the 'old economy'. The golden rule is that you do not say anything, but you simply observe.

Observation techniques are an excellent way to collect objective data, such as frequencies and duration. For example, you could use observation techniques in a 'bricks and mortar' store in your retail sector to get answers to these types of question:

- How long do customers spend selecting goods?
- How long does the checkout process take?
- What percentage of people shop alone?

- How much do customers spend, on average?
- What is the ratio of cash to credit card purchases?

The answers to these questions will help inform the design of the virtual store.

Observation methods are useful for situations when the customer finds it difficult to recall or describe the way he or she carries out the activity. This tends to happen for very frequent tasks or skilled tasks since these can often occur on 'automatic pilot'. Driving is the classic example: try to describe the various actions you execute when turning right at a T-junction (or turning left for readers outside the UK) and you will get some idea of the problem.

An additional advantage of these techniques is that the observer rapidly becomes an expert in the domain area. You do not need to see many customers carrying out the activity before you are able to empathise with their problems and generate imaginative solutions.

However, observing customers also has disadvantages: it is time consuming. It is not really practical to carry out an observation session in much less than a day, and depending on the number of people you have to observe it may take as long as a week. You should also allow an equivalent time for data analysis (as a rule of thumb, each day spent observing takes about a day to analyse). You can save some time by using data recording tools. Various behaviour logging tools exist, or you could always use a pen, paper and stopwatch ...

Video recording

One alternative to being physically present while watching customers is to set up a video camera and record customers' behaviour. You will need permission before you start, but this is potentially a powerful technique to collect data automatically. By using the time stamp on the video camera you can estimate task completion times without needing a stopwatch or software logging tool.

Nevertheless, reviewing tapes can be very time consuming. In addition, no matter how careful you are when setting up your recording equipment you always find that people sometimes move outside of camera shot. Cameras can also make people uncomfortable and this can bias the data you collect.

Interviews and customer visits

Interviews and customer visits are probably the most cost-effective way of getting the necessary background information on your customers. Sampled correctly, interviews with around ten customers will usually get you 80 per cent of the way to a good understanding of your typical customer.

Interviews provide a flexible way to understand your customers' goals and intentions because they allow you to use follow-up questions. Even a short interview will help you get inside the mind of the customer and begin to understand how your customers think about their task.

Interviews are also a good opportunity to sanity-check your assumptions about customers' tasks and goals. Ask customers to comment and elaborate on their tasks; this will provide you with the right terminology and language to use when you express them in an interface. These collaborative sessions help you develop rapport with customers, and this can be useful if you need to return later to ask for clarification.

As with surveys, it is very easy to bias the responses of interviewees with poorly phrased or 'loaded' questions, so you must be careful to appear independent when you are asking questions. It is also a fact that what people *say* may be different to what they actually *do*. For example, people tend to describe their behaviours based on what is socially desirable. So asking someone if they prefer to choose low-fat options in a supermarket will give you a different answer to looking at what products the person has in their shopping trolley.

Customer visits allow you to combine interview techniques with observation of customers in the place where the tasks are carried out. This helps you contextualise the observations that you make and also provides customers with events and artefacts that spark ideas (hence the method is sometimes referred to as 'contextual inquiry'). This provides more valuable data than simply interviewing customers – because when you interview customers they may tell you what they think they are *supposed* to do (rather than what they *actually* do).

Direct methods: summary

DIRECT OBSERVATION

- Advantages: Provides objective data (e.g. frequency, duration); allows identification of important task characteristics; provides information on automatic or forgotten actions; allows observer to develop an understanding of the people doing the work.
- Disadvantages: Time consuming; requires a basic understanding of the task domain; needs data recording tools to be effective.

VIDEO RECORDING

- Advantages: Can take place automatically; provides information on the environment as well as the task; if it includes a time stamp you can use it to estimate time per task; useful supplement to real time observations.
- Disadvantages: Reviewing tapes is time consuming; behaviour can be missed if people move outside of camera range; cameras can make people uncomfortable and so bias the data; some customers may not allow it for security reasons.

INTERVIEWS

- Advantages: Effective tool for understanding goals and intentions, understanding customers' thinking about their work, and their subjective reactions to their work and working environment; good for reviewing first-pass understanding of goals and work processes; can develop rapport with customer.
- Disadvantages: Bias introduced by interviewer; perceptions differ from reality.

An example

Imagine we are developing a new web site that aims to help system administrators ('sysadmins') manage their network with a downloadable tool – a 'network scan and repair tool'. We spend some time visiting four sites of potential customers. At some sites we manage to interview customers; at others we just observe them doing their job. How might we summarise the data we collect?

Table 6.2 shows how we describe the data (in the 'Description' column). The 'source and confidence' column provides an estimate of how sure we are that the data is accurate. For example, the observation that 90 per cent of the customer base is male is based on all the visits; our confidence is therefore high. In contrast, the concern that they do not like documentation only in electronic form was heard at just one site. Our confidence is therefore low – but we do not want to discard this finding since it may later prove to be important.

The form in Appendix 3 also provides a column for describing the design implications of the finding. For example, one way to address the customers' concerns over electronic documentation might be to consider providing a brief 'Getting Started' manual in paper form.

What information do you need?

Both direct and indirect techniques will generate a lot of data. You then need to summarise these data in a form that will help you drive your web site design forward. The purpose of a summary is to prioritise the information that you have collected and present it in a concise form. In Appendix 3 you can find a simple form to help you summarise the results.

Compared with the reams of notes you will have taken during an interview or an observation, this form looks laughably concise. Remember the mantra: it is more important to provide a short summary of the data than aim for accuracy to two decimal places. If this information stays inside your head you have failed. The aim is to make sure other people live this stuff too, and you will never achieve this if you present your audience with a tome of information, no matter how accurate it is. You need to edit ruthlessly.

Table 6.2 An example customer profile

Characteristics	Description	Source and confidence
Personal/physical details	90% male, aged mid-30s. About half wear spectacles. No physical disabilities.	Based on 4 visits; high confidence
Job profile	Responsible for helping users to set up PCs, printers and other network devices on a LAN and fixing any problems that occur.	4 customer visits; high confidence
Education	University education; informal Sysadmins have little formal technical training, most skills learnt on-the-job.	2 customer visits; medium confidence
Domain knowledge	Has knowledge of network protocols esp. TCP/IP, and fixing and debugging network problems.	3 visits. High confidence.
Style preferences	Definite preference for GUI environments, mostly NT but some Macintosh.	2 visits. Medium confidence.
Concerns	Dislike software that has documentation only in electronic form.	Heard at 1 customer site. Low.
Wants	One click installation.	Heard at 2 customer sites; medium confidence

Box 6.1 The six blind men of Indostan

The main issue with customer profiling is making sure that you have full coverage of all user types. It is very tempting to generalise from a small sample. The dangers inherent in this approach are summarised in the fable of the 'Six blind men of Indostan'.

In this fable, a king and his army are passing through the desert near a city in which all of the people are blind. The king has with him an elephant. The people in the city had heard of elephants, but had never had the opportunity to know one. Six young men rush out of the village, eager to find out what the elephant is like. The first young man runs into the side of the elephant. He spreads out his arms and thinks, 'This animal is like a wall'. The second blind man grabs the elephant's trunk. The young man thinks, 'This elephant is like a snake'. The third blind man walks into the elephant's tusk. He thinks, 'The elephant is hard and sharp like a spear'. The fourth young blind man stumbles into one of the elephant's legs. He thinks, 'An elephant is like a tree trunk or a mighty column'. The fifth young blind man finds the elephant's tail and thinks, 'The elephant is nothing but a frayed bit of rope'. The sixth young blind man grasps the elephant's ear and feels its thin roughness. He thinks, 'The elephant is like a living fan'.

Finally, an old blind man comes. He had left the city, walking in his usual slow way, content to take his time and study the elephant thoroughly. He walks all around the elephant, touching every part of it, smelling it, listening to all of its sounds. He finds the elephant's mouth and feeds it a treat, then pets it on its great trunk. Finally he returns to the city, only to find the city in uproar. Each of the six young men had acquired followers who eagerly heard his story. But then, as the people found that there were six different contradictory descriptions, they all began to argue. The old man quietly listened to the fighting. 'It's like a wall!' 'No, it's like a snake!' 'No, it's like a spear!' 'No, it's like a tree!' 'No, it's like a rope!' 'No, it's like a fan!' The old man turned and went home, laughing as he remembered his own foolishness as a young man. Like these, he once hastily concluded that he understood the whole of something when he had experienced only a part.

You might find this story consoling next time you find yourself in a design meeting with people arguing over what the user is 'really' like.

Other potential pitfalls include paying too much attention to:

- the last customer visited;
- the customer who complains the loudest;
- the developer who claims to know exactly what customers want.

The key areas to summarise are provided below:

Personal and physical details

- Name
- Contact details (for later follow-up)
- Native language
- Age
- Sex
- Likely problems with input device (mouse/keyboard, etc.) or reading screen
- Physical limitations.

Web site behaviour

- Why would they visit the site?
- How do they find web sites (search engines, portals)?
- Expectations of site behaviour (conventional or unconventional)
- Preferred navigation style (hierarchical/linear)
- Attitude to innovative user interface styles (for example, Flash).

Education

- Highest qualification achieved
- Key topics studied
- Developed applicable skills
- Reading level and comprehension
- Typing speed.

Domain knowledge/experience

- Everyday concepts used when describing or using product
- Experience with current web site
- Experience with competitor web sites (which ones?)
- Experience with web sites that have similar functions
- Familiarity with computer systems in general
- Experience with specific applications, interfaces or operating systems (which ones?).

Style preferences

- Intellectual styles (approach to problem solving)
- Working styles
- Learning styles
- Preferred writing tone

- Attitudes to product
- Attitudes to IT
- Attitudes to brand
- Life values
- Discretion to use web site.

Concerns and wants

- Current problems with carrying out these tasks
- Key trends in this area
- Aspects of task currently found unrewarding and would like to disappear/ diminish
- What they expressly don't want to see in a new web site
- Key enhancements they want to see
- Aspects of task found rewarding and would like to see more of.

How to present the information

Since customer information is such an important driver for the design of the site, it makes sense to present the information in ways that engage the design team. One technique that has proven to be successful is the production of customer 'personas'.

Personas are short, engaging summaries of the key customer groups. They describe customer archetypes – neither an 'average' nor a real customer, but a stereotypical one. Personas aim to bring the customer to life, so the personas are given a name and usually include:

- a photograph (use clip art or a picture from a trade magazine);
- a quotation that captures the customer's key objective;
- a short narrative describing the customer;
- a list of the customer's key goals;
- a status designation to capture the customer's importance in the overall project (for example, primary or secondary customer, or even a 'negative' customer: one that the site is specifically *not* aimed at).

The key benefit of this approach is that the design team can then discuss features and functionality with direct reference to 'real' people: in contrast to having a single, anonymous and 'elastic' user that can be stretched to justify almost any design decision. The rationale is that, by designing for a prime example of the customer group – a 'somebody' – you have a very clear understanding of that customer's behaviour, goals and motivations. By designing 'everything for somebody', you will then satisfy the broader group of somebodies represented by that prime example.

For detailed background information on personas, the best source is Alan Cooper's *The Inmates are Running the Asylum* (see the further reading section at the end of this chapter). But the concept behind personas is really quite simple and can be well communicated with examples. Imagine that we are designing a web site aimed at people who want to pick shares. Imagine that our web site will allow our customers to track a portfolio of shares, search for shares that meet particular criteria, and provide analysis tools to rate the performance of shares on various financial statistics.

After carrying out detailed interviews, we identify three key customer types. The first type of customer is fairly new to share dealing and simply wants to know how his or her collection of shares is doing. The second type of customer wants to carry out detailed research on a range of companies and make his or her own judgements on where to invest. The third type of customer wants to trade in shares actively but prefers to choose companies based on the recommendations of experts.

Interviews and research into these three types of customer result in the construction of the personas shown in Figure 6.2.

You can see from these examples that personas summarise the key information collected in the customer profiling exercise. Note that you may not have interviewed a customer who is *exactly* like this. The customers that you use in personas are composite customers or stereotypes.

Even though the personas are extremely brief, they still contain enough information to make high-level judgements on whether or not particular features should be developed – and in particular, which customer archetype will benefit. When it comes to the development itself, the design team will begin to ask more detailed questions that are not answered in the persona. For example, one developer may want to know if David prefers a visual style of data presentation or if he prefers to have the data in a downloadable spreadsheet that he can then analyse in detail off-line. To answer this question you will need either to refer to your customer profiling form or write additional, more detailed personas that contain some of this ancillary information. One technique that has worked well in my own consulting assignments is to film scripted interviews with the persona (an actor plays the role of the persona). These can be shown in development meetings and left on the company's intranet for later viewing.

The next step is to identify the 'primary' persona: the customer who will be the main focus of the design work. With most projects, this can usually be agreed after a discussion with the project stakeholders. For example, in the personas above our research may suggest that David is the primary persona. If no decision can be made on the primary persona, this suggests that there are two or more equally important customer groups. The implication of this is that each customer group requires its own interface.

The aim is to help the design team empathise with the personas. So make them as real as possible – print them out on posters, hang them around the office and insist that discussions of new features are firmly discussed by reference to the personas you have identified.

'Am I still on track to retire at 55?'

Cautious Colin

Colin Wright

Colin is 42 and works for a large blue-chip company. He tops up his company pension by investing in safe, secure shares. He invests in new shares once or twice a year and tends to hold them even when they drop in value.

COLIN'S GOALS

- Manage finances in minimum time
- Achieve financial security through steady investment
- Avoid complexity
- Use technology as a tool.

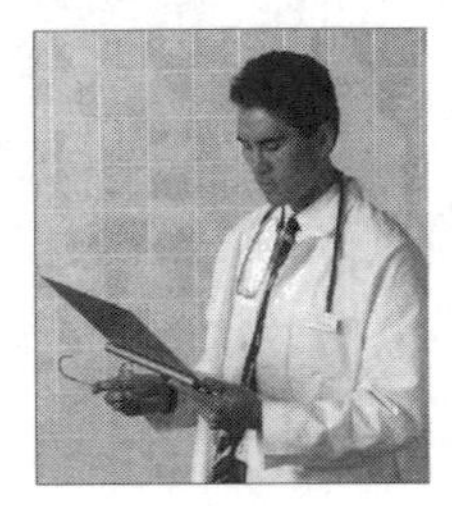

'I think it's time to move out of biotech stocks.'

Data-driven David

David Armstrong

David is 35 and works as a doctor in a large city hospital. He likes to invest in particular industrial sectors. He tends to invest for the long term, but will sell shares in a company if its management underperforms. Before investing, he carries out extensive research on a company.

DAVID'S GOALS

- Reduce risk through accurate information
- See the big picture, including the gaps
- Find out where to go to get more detail
- Invest ethically.

'Who's up and who's down today?'

Share-dealing Stewart

Stewart Clapham

Stewart is 38 and works for an IT company. He trades shares frequently, often two to three times per week. He is always looking for rising stars and invests either on recommendation or on a hunch.

STEWART'S GOALS

- Make money fast
- Quickly spot investment rockets and falling stars
- Use his understanding of technology to gain an edge
- Get timely information.

Figure 6.2 Three example personas

Summary

- To achieve a customer-centred web site you need to understand something about your customers.
- Indirect methods allow you to get information from your customers when you cannot meet them face-to-face. Typical techniques include surveys, web site logs and artefact analysis.
- Use direct methods when you can actually see customers in action. Typical techniques include observation, video recording, interviews and customer visits.
- There are a number of pitfalls ready to befall the researcher collecting customer data. In particular, be sure you have full coverage before making generalisations.
- Engage the design team with the data by expressing the results in the form of customer personas.

Further reading and weblinks

Books and articles

Cooper, A. (1999) *The Inmates are Running the Asylum*. Indianapolis, IN: SAMS.
Hackos, J.T. and Redish, J.C. (1998) *User and Task Analysis for Interface Design*. New York: John Wiley & Sons.
Mayhew, D.J. (1999) *The Usability Engineering Lifecycle*. San Francisco, CA: Morgan Kaufman Publishers.
Nielsen, J. (1993) *Usability Engineering*. Boston, MA: Academic Press.

Web pages

Cooper Interaction Design (2002) 'Concept projects'. http://www.cooper.com/concept_projects.htm. This article shows how personas are used in the context of a design project (a design concept for a PDA application to help people find their way around an airport).
Fuccella, J., Pizzolato, J., Franks, J. (1998) 'Web site user centered design: techniques for gathering requirements and tasks'. http://internettg.org/newsletter/june98/user_requirements.html. 'The intent of this article is to provide Web site usability engineers with practical techniques for quickly and effectively identifying user expectations for their sites, both in terms of content and functionality.'
Goodwin, K. (2001) 'Perfecting your personas'. http://www.cooper.com/newsletters/2001_07/perfecting_your_personas.htm. 'These tips will help you refine your personas so you can get the most out of them.'
McDunn, R.A. (2001) 'Web server log file analysis – basics'. http://www-group.slac.stanford.edu/techpubs/logfiles/info.html. This article provides a tutorial for analysing your web server access logs.
Nielsen, J. (2001) 'Are users stupid?'. http://www.useit.com/alertbox/20010204.html. 'Opponents of the usability movement claim that it focuses on stupid users and that most users can easily overcome complexity. In reality, even smart users prefer pursuing their own goals to navigating idiosyncratic designs. As Web use grows, the price of ignoring usability will only increase.'

Venn, J. (2001) 'Stalk your user'. http://www.webtechniques.com/archives/2001/06/veen/. 'Design, ultimately, is problem solving. And the best way to discover which problems need solving is to look for them in context.'

7 Build environment profiles

The second part of the context of use jigsaw is the environment profile. An environment profile describes some of the obvious physical characteristics of the environment (such as noise and lighting levels), but it also describes some of the less obvious environmental constraints, such as the socio-cultural and technical environments.

Even for web users, the environment can vary significantly. Familiar examples include screen resolution and the type of browser. It is common for designers to design for a target screen resolution of 800 × 600, even though many customers will be using screen resolutions smaller (such as 640 × 480) and larger (such as 1024 × 768) than this. Other customers may be working from a different screen device entirely, such as a mobile device. Screen resolution has very significant implications for the way information is presented and formatted, but it also has implications for the type of content that is presented. Mobile users will not welcome large graphics and they will also want short, information-rich content.

To adequately characterise the environment in which your site will be used, you will need to summarise information about the physical environment, the socio-cultural environment and the technical environment.

Physical environment

The physical environment refers to issues such as:

- What does the workspace look like? What is the size of the workspace: cramped or generous? Will the customer be sitting at a desk in front of a computer? Will the customer be on the move (for example with a PDA or WAP device)? Take pictures if possible.
- Light levels: will office lighting be the norm or might the web site be used in low light levels?
- Noise levels: will the customer be accessing the site from a quiet office? From home, with music playing or with the television on? On the move, in traffic, on a train?

- Thermal environment: for example, if this is a public kiosk system the customer may be wearing gloves. What might be the implications of this for using a keyboard?
- Visual environment: what other parts of the visual environment might distract the customer from his or her task (for example, imagine a web browser within an interactive TV system that has a picture-in-picture view of the current TV programme)?
- Location relative to other people and equipment: for example, if this is an on-line banking application the customer may not want people peering at the screen.
- Location and availability of any equipment or information to support task: for example, credit cards, delivery addresses ...
- Customer posture or position: lean forward (PC use) or lean back (interactive TV use)?

Socio-cultural environment

Whether your customers access your site from home or from work, they do so in a socio-cultural context. Neither of these environments is isolated from the world outside and both are part of a wider community that has cultural practices and social norms. Rituals and customs are common in all cultures and web sites are used within the context of these.

A couple of examples might make this clear:

- A customer searching for a family holiday at your package holiday web site is unlikely to buy a holiday on the first visit. It is more likely that he or she will want to print off information about two or three holidays and discuss the options with the family before making a decision. This has profound implications about functionality: your site should allow customers to compare the options on various holidays, pages should be formatted for printing, and returning customers will want to get back to where they left off.
- A business user visiting your site may be working in an environment where access to the Internet is considered a privilege, or thought of as 'not real work'. This has implications for the visual messages that your site sends out to other people working in the office. Put simply, in this context the customer will want the site to look functional and business-like and may not want the site to look 'fun'.

The list below provides some examples of issues you should address when examining the socio-cultural environment. (Note that some items in this list are specific to business-to-business sites and so refer to the environment in the customer's workplace.)

- Task practices
- Team working

- Management structure
- Organisational aims
- Communication structure
- Sharing
- Security/privacy
- Assistance required or available
- Policy on web use
- Job flexibility
- Autonomy
- Hours of work
- Interruption
- Breaks
- Performance monitoring
- Performance feedback
- Pacing.

Technical environment

Clearly, the customer's technical environment needs to be sufficient to access your site. If customers do not have the appropriate plug-in, your site may as well not exist for them.

The technical environment refers to issues such as:

- Hardware
- Mobile or fixed system
- Networked or stand alone
- Screen resolution
- Colour depth
- Processor (e.g. Wintel)
- Software
- Custom or off-the-shelf software
- Type of operating system (e.g. Windows, Mac OS, Palm, WAP)
- Browser type and version
- Reference materials
- Manuals
- On-line help
- 'Quick tips' card.

Statistics on some of these issues are widely available. For example, at the time of writing (April 2002), most customers (that is, half or more) use Microsoft Internet Explorer 5.x, use a full-colour (16 bit) 800 × 600 resolution screen, and use the Microsoft Windows 98 operating system. More detailed statistics are given in Tables 7.1 to 7.4. Be aware that these tables provide statistics on the 'average' web user: as we discuss throughout the book, your customers may not be average and you should sanity check these figures against data for your own site. (Note that the percentages in these tables do

Table 7.1 Screen resolutions (from TheCounter.com)

Screen resolution	*Percentage of web users*
800 × 600	51
1024 × 768	35
640 × 480	3
1280 × 1024	3
1152 × 864	3
1600 × 1200	< 1

Table 7.2 Colour depth (from TheCounter.com)

Colour depth	*Percentage of web users*
Thousands (16 bit)	50
Billions (32 bit)	34
Millions (24 bit)	10
256	4
16	< 1

Table 7.3 Operating systems (from TheCounter.com)

Operating system	*Percentage of web users*
Windows 98	58
Windows 2000	17
Windows ME	7
Windows 95	5
Macintosh	2
Windows 3.x	< 1
WebTV	< 1
Linux	< 1
Unix	< 1
Windows XP	< 1
OS/2	< 1
Amiga	< 1

Table 7.4 Browser types (from TheCounter.com)

Browser type	*Percentage of web users*
MSIE 5.x	56
MSIE 6.x	32
Netscape 4.x	4
MSIE 4.x	3
Netscape compatible	1
Netscape 6.x	< 1
Opera x.x	< 1
Netscape 5.x	< 1
Netscape 3.x	< 1
MSIE 2.x	< 1
MSIE 3.x	< 1
Netscape 2.x	< 1
Netscape 1.x	< 1
Amiga	< 1

Table 7.5 Example of environment summary

Characteristics	Description
Physical environment	*Office environment*: lighting supplied by overhead fluorescent, around 100 lux. Temperature between 65–75°F. *Home environment*: Small office or kitchen table.
Socio-cultural environment	*Office users*: 58 per cent access the web exclusively or primarily from work. Some of these users are allowed to use the web for personal use during lunch hours. *Home users*: access the web primarily after 6pm and at week-ends. Users tend to surf alone. Large purchase decisions are first discussed with partners or spouses.
Technical environment	We found significant differences compared to the 'average' Internet user defined at TheCounter.com. In our customer base, 82 per cent use Microsoft platform, 13 per cent use Macintosh. About 70 per cent connect via 56 k modems or slower. 30 per cent have high bandwidth connections (>1 Mb/sec). 75 per cent of users use Internet Explorer with 25 per cent using Netscape (Netscape users tend to be 'old timers' who have used the web for many years). 80 per cent of users have used JavaScript. 63 per cent own a colour printer.

not always add up to 100 per cent because, for a small number of users, the statistic cannot be measured.)

How to present the information

Once you have used the lists above to generate the environment profile, you should summarise your information using the form in Appendix 4. An example is provided in Table 7.5.

Summary

- To achieve a customer-centred web site you need to understand the environment in which people will access your site.
- The environment includes the physical environment (e.g. light levels), the socio-cultural environment (e.g. sharing of information) and the technical environment (e.g. browser version).

Further reading and weblinks

Books and articles

Salvendy, G. (ed.) (1997) *Handbook of Human Factors*, 2nd edition. New York: John Wiley & Sons.

Web pages

TheCounter.com (2002) http://www.TheCounter.com/stats. Find out how Internet users are viewing the Web (monthly statistics on browsers, OS, screen resolution, etc.).

8 Build task profiles

Task profiling is one of the areas most likely to be overlooked when developing a web site. Most web producers or site managers will be able to hazard a guess about likely customers. They will also have some definite assumptions about the technical environment, such as the browser or screen resolution. But now ask them what they expect customers to do at their site and answers tend to be *post hoc* and couched in terms of implementation. For example, most sites on the web have a search engine, so it is easy to assert that your own site needs a search engine. You can then generate a *post hoc* task: customers will want to search. In fact, customers may not want to search – customers *will* want to find information or products at your site, but 'Search' is just one way of achieving this goal. This is not a semantic argument: for your customers, a better implementation may be a category listing or a site map or an alphabetical index of pages. Similarly, registration screens, chat rooms and streaming audio are ways of implementing particular functionality that may or may not be important to your customers.

Logically, if we are going to design a site that is focused on customers' tasks, we must have a very clear idea about the specific tasks that customers want to engage in. Without this knowledge we will not be able to design our site to help people complete these tasks. As we said in Chapter 5, something for everybody equals everything for nobody. So it is just as important to say what tasks your site will *not* support.

If you are new to customer-centred design, this is the stage where you will begin to hesitate. This is partly because you do not have enough information. You may not know what your customers want to do and you do not want to hedge them in. You may already have the code to implement a search engine; so how much harm will it cause to include it? After all, it's just a two-centimetre box at the top right of the page – it's not as if it takes up a lot of room. So why not include it anyway?

The reason for insisting every item of functionality on your screen can justify its presence is because each item represents a chance for your customers to make a mistake. For example, if you have to make a choice from amongst two equally-likely alternatives, the chances of you making an error are 2:1.

Increase the number of items to five and your chances reduce to 5:1. Even if each of the items are unambiguous and you are sure that customers will never make the wrong choice it is still a fact that customers will take longer to choose one item from five than one item from two: choice reaction time is a function of the amount of information. There aren't many laws in psychology, but this is one of them (known as 'Hick's Law') so it is worth paying attention to. This means you need to make every navigation item justify its place on your screen.

Some uncomfortable consequences will follow this analysis. For example, you will find that some navigation items on your site are not really needed: these items are not related to customers' key tasks and their only purpose is to add visual noise. It gets worse – your CEO may have specified that some of these items need to be on the home page. Be prepared for a rough ride.

Task profiling and staged delivery

Staged delivery is a common theme in modern software development. The basic idea is very simple: deliver the software is stages, with the most important functionality delivered first. Note the mention of 'the most important functionality'. This means some planning has gone into the decision; it does not mean uploading the minimum web site that you can complete with the time and resources you have.

Important functionality occurs at the intersection of the customers' needs with the company's commercial requirements. Customer centred design in general, and task profiling in particular, offers an ideal framework for defining this functionality. Consider the diagram in Figure 8.1. This diagram plots some of the typical tasks that customers will carry out at your web site. (For simplicity, the schematic shows only a subset of all the likely tasks.) These are labelled Task A through to Task I.

The size of each task 'bubble' relates to the number of customers who will carry out that task. So Task G is a task that will be carried out by a relatively large proportion of customers, whereas comparatively fewer customers will carry out Task A. The vertical axis defines the importance of each of these tasks to customers. So for example, Task B is a highly valued task, whereas Task D is less highly valued. Finally, the horizontal axis represents the ease of implementation. This refers to the amount of software development needed to support each task end-to-end. Task G is amongst the easiest and Task A is amongst the hardest.

We can now use this type of diagram to make educated decisions on what should be in the various stages of our 'staged delivery'. For example, it makes sense to support Tasks G and B in the first release, since these are relatively easy to implement and of high value to customers. In the second release we will add some of the additional tasks that are still highly valued by customers but are harder to implement (such as Task H), or that are of lower value but easier to implement (such as Task E).

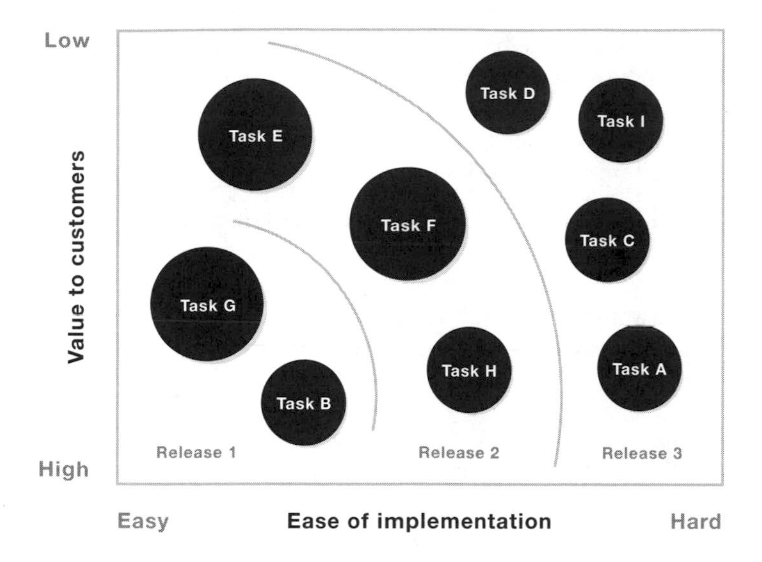

Figure 8.1 Web site tasks – value to customers v. ease of implementation

Develop a task list

The first stage in understanding your customers' tasks is to brainstorm a task list. At this point, you will find that the information collected during customer interviews and customer observations is a great source of ideas. Competitor sites will usually provide you with some ideas too. You should also call on your colleagues, especially those that have some experience with customers. As with all brainstorming sessions, avoid comment, criticism, judgement or evaluation and remember the aim is quantity not quality, so try to generate as many ideas as possible.

The next step is to group related ideas together, so it sometimes helps to write the ideas on sticky notes so they can be moved around on a board.

If you are still having problems generating a comprehensive list, look at any requirements documents that have been produced. Try to read between the lines to understand what tasks many of these requirements have been selected to support. (Although beware: some of the requirements may be wrong.)

As a specific example, assume we are designing a web site for an English football club. What do people want to do at our site? A selection from our brainstorming session might look like this:

- contact the players;
- find out the top goalscorer;
- get travel details for the next away match;
- buy a replica kit;

- read match statistics;
- join the supporters club;
- read a match report for the last game;
- download a fixture list;
- buy a ticket for the next away game;
- get detailed information on a particular player.

Note that this is just a small sample: your task list may comprise 30 or so tasks.

Once the ideas begin to dry up, start grouping tasks together in a way that is logical to customers. Now you can apply some analysis to your list. Some tasks will be duplicates, for example, and some may be just plain wacky (too wacky to support).

Prioritise the list

The next step is to assign some priority to items in your list. We need to do this so that we can identify the most important tasks from the customer's perspective and from the point of view of the organisation. We need to identify the frequent tasks and distinguish them from tasks that are rare. Frequent tasks are crucial to the success of the web site since they will determine customers' perception of it. We also need to distinguish important tasks from less important tasks. Important tasks may be infrequent but customers will hate your web site if these tasks are not well supported. One example of an infrequent but important task is where customers need to install a plug-in before using the site, since failure means they cannot use the system.

Figure 8.2 provides a simple way for you to use this approach to quickly assign a priority to a task.

First, assign a frequency to the task. For example, if the task is carried out almost every time the customer visits your site, or many times during a single visit, the frequency is high. An example of a high frequency task might include placing an order. In contrast, you should assign a low frequency to tasks carried out rarely. Typical examples might include registration or reading the company's annual report to shareholders.

Next, define the task's importance. For example, placing an order is a critical task, since, if the customer fails, the site will not make money. We

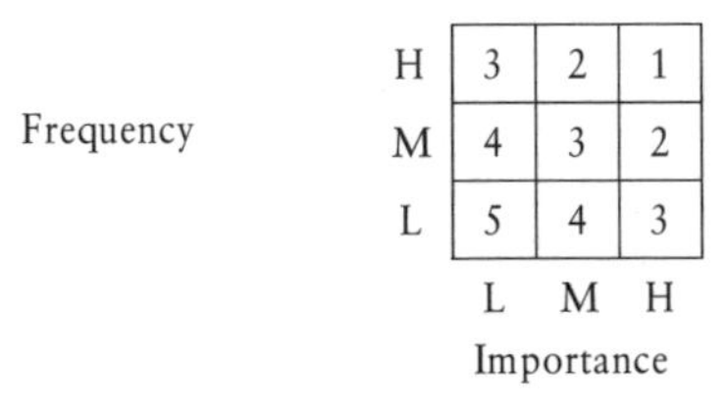

Figure 8.2 A method of assigning the priority to a task based on frequency and importance

Table 8.1 An example of assigning priorities to four different tasks

Task	*Frequency*	*Importance*	*Priority*
Logging on	High	High	1
Using search engine	High	Medium	2
Registration	Low	High	3
Reading Annual Report	Low	Low	5

therefore define the importance as high. In comparison, reading the annual report we might define as low since a paper copy will have been sent to all shareholders. However, when making a judgement on importance, remember that this is a commercial decision too. So for example, registration may be of low importance to customers but highly significant to the organisation.

Use Figure 8.2 to assign a priority to each task. Table 8.1 provides an example.

The table (reproduced in Appendix 5) can then be used to provide guidance in identifying the high priority tasks.

Get detailed information on each task

Now we have a prioritised list of tasks, we need to start collecting more detailed information for each of them. So for each of the tasks in your task list, write down the information under the following headings. A form for this purpose is provided in Appendix 6.

Goal of task

- Why is this task performed?
- What higher level task is it part of?
- What is the business context for this task?
- What are the criteria for success?

People involved

- Which customer performs the task?
- What skills are required to be successful in the task?
- What other interactions with people are there?

Problems and difficulties

- Is the task simple or difficult?
- Are there opportunities for errors or problems with the task?
- What are the consequences of errors?

Time

- How frequently is the task performed?
- How long does the task take?
- Does the customer have any flexibility?
- Are there any time constraints on the task?
- Is any pacing imposed or necessary?

Inputs and outputs

- What is the starting point or input?
- Where does any necessary input come from?
- What is the end point or output?
- Where does the output go?
- What information is required to complete the task successfully?

Results

- What results are required (collect examples/artefacts where possible)?
- What quality levels are expected?
- How are the results to be used?

How to present the information

In Chapter 6, we used personas to help us describe the customers of the web site to the design team. You will remember that personas are short, engaging summaries of stereotypical customers. Personas summarised the information that we collected during the interviews with customers.

We need a similar technique to describe the customers' tasks, and scenarios are a practical way to achieve this. Scenarios are realistic stories that describe the use of the web site from the customer's perspective. These scenarios summarise all of the detailed information you have collected in a short narrative. They describe motivations for the customer's actions and include a description of the customer's overall goal in using the site.

In Chapter 6 we discussed the design of a web site aimed at people who want to pick shares. This web site will allow customers to track a portfolio of shares, search for shares that meet particular criteria, and provide analysis tools to rate the performance of shares on various financial statistics. How might a task scenario look for this site?

Our first step would have been to brainstorm a task list. Then we would have prioritised the list and next we would have obtained more detail on each of these tasks. To write the scenario we then look at the task list and see if any of these tasks logically go together. For example, if this was an on-line banking site, then the task 'Check balance' is likely to be linked to some other tasks such as 'Pay a bill'. In this way, we chain together the tasks to represent a realistic, end-to-end interaction with the web site.

Box 8.1 Can we always identify 'tasks' on the web?

This discussion of 'tasks' implies that customers are very goal-oriented when using web sites. What about sites that do not expect users to achieve tasks – for example, sites that want to entertain users?

Virtually all of the research in this area does indeed support the notion that customers are very goal-oriented when using the web: they actively try to achieve specific tasks. As far as e-commerce sites are concerned, this should not surprise us: most people carrying out transactions in the real world act the same way. For example, the person looking at brochures in a travel agency is probably seeking out a holiday. You could turn this into a 'fun' browsing experience by hiding some of the brochures in odd locations, or arranging the brochures in an artistic manner and seeing what happens, but the chances are that this will lose you customers since the aim is not to be entertained but to be find a holiday.

This does not mean that customers simply want a functional experience: certainly, customers want to be engaged. To extend the analogy, a travel agent that makes suggestions for new holidays based on asking questions about my previous holidays will engage me in ways that simply providing access to brochures will not. But the key to the engaging experience is to provide ways to help me achieve my goal in a faster or a more effective or a more satisfying manner.

For our share dealing web site, we might chain together the following tasks:

- sell shares in a company;
- buy shares in a company;
- view a list of 'ethical' investments;
- carry out research on a company;
- look at the performance of a company over a number of years;
- compare a group of companies on financial statistics.

The aim of our scenario is to combine these tasks into a realistic narrative. An example might be the following scenario, written for 'Data-driven David':

> David has read in the newspaper that RxDrugs, a pharmaceuticals company, are developing a new anti-depressant. Although he invests in this particular sector, he does not have any shares in RxDrugs and decides to use the web site to find out some more about the company. He does not have any new money to invest, so if he decides to put money into RxDrugs he will need to sell some of his existing holdings. To help him make this decision, he wants to compare RxDrugs with the other companies on a range of financial statistics, including growth, P/E ratio

and earnings per share. He wants to see these statistics over the past five years. David will also want to check that RxDrugs meets his criteria for ethical investment.

All of David's research convinces him that RxDrugs is a worthwhile investment. However, he likes to have a 'cooling off' period, and decides to consider the decision overnight.

The next day he returns to the site, and sells sufficient of his shares in another pharmaceuticals company to invest £4,500 in RxDrugs.

Note that this scenario describes a realistic end-to-end task. It is also tied to a particular persona – it is not David's style to buy shares immediately but to consider his decision overnight. This has profound implications for the design of the site: to support this task, the web site may need to have some kind of 'memory' so that David can quickly return to the point he was at the day before (although note that the scenario places no requirement on the implementation: it could be achieved by cookies, user authentication or some other method). If, instead, we were focusing on 'Share-dealing Stewart', the scenario would need to include an immediate purchase.

Useful questions to bear in mind when reviewing your scenarios include:

- Is the scenario accurate and realistic?
- Is the scenario specific and measurable?
- Does the scenario describe a complete job (integrated, not simple tasks)?
- Does the scenario describe what the customer wants to do (not how the customer will do it)?
- Is the scenario 'portable' to equivalent web sites?

Appendix 7 provides a form for developing your own scenarios.

Like personas, scenarios are a powerful and persuasive tool. They help the design team distinguish important *functions* from less important *features*. They provide a consensus on precisely what customers are trying to do with the site. And because they are written as accessible narratives, rather than in arcane notation, they allow all people on the design team – marketing, development, customer support – to contribute equally to the design.

Scenarios are also useful beyond the initial design stage. With slight adaptation, they can become the scripts that are used in usability tests. A customer is provided with a prototype of the web site and a copy of the scenario and then asked to complete the task. This is an area we will return to in the next part of the book as we begin to create the customer experience.

Summary

- Task profiling provides you with a description of what customers will do at your site.
- Task profiling is a key element of staged delivery: providing the right features in the right way at the right time.

- Begin by brainstorming a task list, then prioritise the list and collect detailed information on each of the tasks.
- Summarise the data you collect by describing the tasks as scenarios. Scenarios are end-to-end descriptions of realistic interactions with the web site, expressed from the customer's perspective.

Further reading and weblinks

Books and papers

Hackos, J.T. and Redish, J.C. (1998) *User and Task Analysis for Interface Design*. New York: John Wiley & Sons.

Nielsen, J. (1993) *Usability Engineering*. Boston, MA: Academic Press.

Rosson, M.B. and Carroll, J.M. (2001) *Usability Engineering: Scenario-Based Development of Human Computer Interaction*. London: Morgan Kaufmann Publishers.

Web pages

Travis, D.S. (1997) 'When GUIs fail'. http://www.system-concepts.com/articles/gui.html. 'Usable systems are defined by their focus on the user's tasks, not by a pretty interface. Our informal studies show that 60 per cent of a system's usability comes from task focus, 25 per cent from consistency and just 15 per cent from presentation of information.'

Travis, D.S. (1997) 'How to turn your dot.com into a hot.com'. http://www.system-concepts.com/articles/experience.html. 'Experience scenarios help you get to the essence of what customers want to do at your site. They differ from simple use cases, because use cases rarely include context information and tend to focus on atomic tasks.'

Lewis, C. and Rieman, J. (1993) 'Task-centered user interface design. Chapter 2: getting to know users and their tasks'. http://www.hcibib.org/tcuid/chap-2.html. Every chapter in this $5 shareware book is worth reading, but this chapter on users and tasks is especially good.

Part III
Step 3
Create the user experience

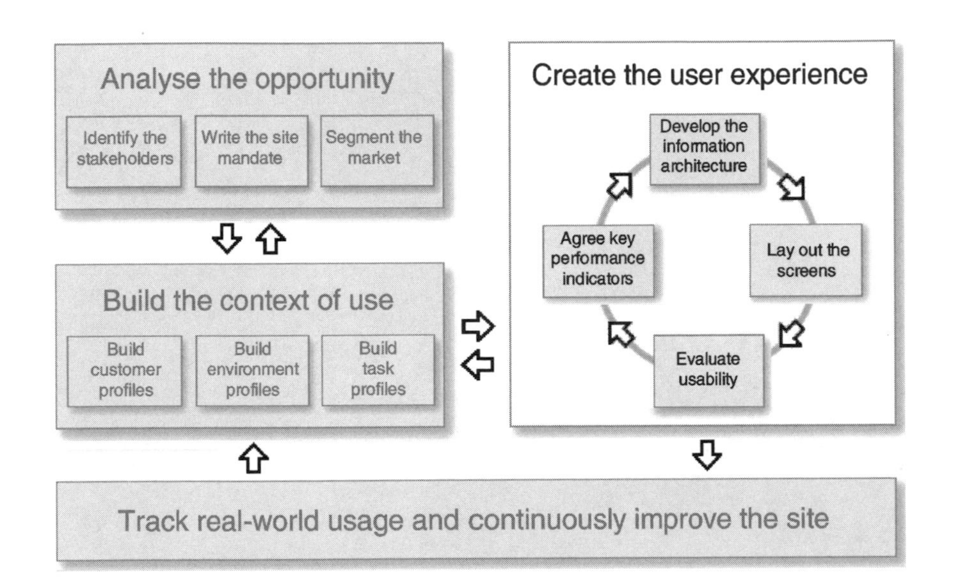

9 Agree key performance indicators

Usability is perceived by many to be a 'soft' attribute that is difficult to measure. Certainly, if we characterise usability as simply a feeling or an attitude that a customer has to a site, then it will be difficult to quantify. Think of a web site – or indeed any piece of software – that you feel is 'usable'. What makes this system usable?

It is probably one or more of the following:

- Useful: it does something that is hard to do in any other way
- Flexible: it allows you to solve a problem in more than one way
- Memorable: it is easy to remember how to do most things
- Quick to learn: most people can get going with the system fairly quickly
- Fun to use: using the product is fun in its own right
- Attractive: it is engaging to look at
- Efficient: it helps you get things done more quickly or more cheaply
- Simple: the most important things can be done in one or two steps.

Although some of these attributes – such as 'simple' – seem 'soft' at first, each of them can be broken down, clarified and eventually quantified. The aim of this chapter is to help you specify the attributes that are critical for your site and then show you how to quantify the attribute and measure it.

We achieve this by specifying 'key performance indicators' for the web site. Key performance indicators are usability benchmarks – quality measures – that you use to say if the site is ready to release. They are clearly defined, measurable attributes of a system.

Components of key performance indicators

Each key performance indicator has four components:

1. High-level objective
2. Context of use
3. Measurement technique
4. Performance values.

Each of these steps aims for a progressively finer level of definition, as you will see.

High-level objective

The high-level objective is an aspiration statement that answers the question: 'What needs to happen for the site to be considered a success?'

As a specific example, imagine we are working on the development of an on-line bank. A suitable high-level objective might be: 'Reduce costs by allowing customers to fully manage their account on-line'. This objective makes the commercial objectives clear ('reduce costs') as well as the benefit to customers ('fully manage their account'). It also describes the customer's experience with the complete site and not just one component of it. We then enter this objective in the key performance indicators form (Appendix 8); Figure 9.1 shows an example.

Here are some more examples that might apply to different sites:

- 'Increase our market share by being perceived as the fastest one-stop-shop for share dealing'
- 'Increase revenue by eliminating the costs of our call centres and by providing 24/7 access'
- 'Position our brand as "easy to do business with".'

You should avoid anodyne objectives, such as 'make the site easy to use' or 'keep our customers happy' since these are so general that they are effectively meaningless.

Context of use

The second component of the key performance indicator is the context of use. This describes a specific customer carrying out a specific task under specific circumstances:

- Who?
- Doing what?
- Under what circumstances?

Key performance indicator for: On-Line Bank

High-level Objective
Reduce costs by allowing customers to fully manage their account on-line.

Figure 9.1 Key performance indicator: high-level objective for an on-line bank

Context of use	
Who?	Experienced web users
Doing what?	Managing their bank account
Under what circumstances?	Using a PC at home and work

Figure 9.2 Key performance indicator: context of use for an on-line bank

To continue with the on-line banking example, we might describe the context of use as 'Experienced web users managing their bank account using a PC at home and work'. The second part of the template helps us to generate this statement. Figure 9.2 shows an example.

Some other examples:

- '50–60 yr old first-time web users getting a quotation for car insurance using an interactive television'.
- 'A customer with two-years' share dealing experience deciding which telecommunications shares to add to an ISA using a PC at home'.

For some criteria, you may question if a customer-centred measurement is really necessary. For example, if your key performance indicator relates to the speed of the site, why not just measure the speed with an automated tool, or sit there with a stopwatch and measure the amount of time a page takes to load?

The reason for this is that it is only the customer's perception of speed that matters. The page may download five seconds more quickly than a competitor site, but if customers *perceive* it to be slower then it is slower. This is not a question of semantics, but a real, measurable effect. So for example if a page takes 20 seconds to download, but the page loads progressively in the browser window, customers will think this is quicker than a page that takes 15 seconds to load if they cannot see anything until the page has been fully loaded into the computer memory. Researchers have also shown that the perceived speed of a web site relates to how quickly people can complete their tasks, not how quickly pages download (for more detail, see the Further Reading section at the end of this chapter).

Measurement technique

The third step is to refine this further: we now describe the scale we will use for the measurement and how we will actually derive the values.

The customer centred scales that we use should be related either to customer performance or to customer perception. At this point, it is worth considering the definition of usability given in the International usability standard, ISO 9241-11:

> Extent to which a product can be used by specified users to achieve specified goals with effectiveness, efficiency and satisfaction in a specified context of use.

Note the three components in this definition:

- Effectiveness: the accuracy and completeness with which customers achieve specified goals.
- Efficiency: the accuracy and completeness of goals achieved in relation to resources.
- Satisfaction: freedom from discomfort, and positive attitudes towards the use of the web site.

This definition provides us with more precise and objective terms with which to operationalise and specify usability. It also provides us with terms with which we can assign numerical values. With this approach to evaluating usability, the assessment of our web site is couched strictly in terms of customer performance and satisfaction where the customer is the critical part of the web site. This is in contrast to other forms of evaluation, such as functional testing, which usually tests the system in isolation from end-user involvement.

It is important to note that performance and attitude measures are often independent. That is, a web site may be effective and efficient but the customer may have a poor opinion of it. Therefore all three aspects of usability must be addressed in the design, development and evaluation of the site in order to optimise the customer experience.

Measures of effectiveness

Effectiveness means the accuracy and completeness with which customers achieve specified goals. Probably the most common measure of effectiveness taken by usability practitioners is the completion rate: the percentage of customers who successfully complete the task or tasks.

Other examples of metrics that you can use for this include:

- number of power tasks performed;
- percentage of relevant functions used;
- percentage of tasks completed successfully on first attempt;
- amount of the task completed successfully;
- number of persistent errors;
- number of errors per unit of time;
- percentage of customers able to successfully complete the task;

Table 9.1 Simulated task completion data from a usability test (in percentages)

Task	*Search*	*Price*	*Buy*
P1	100	50	50
P2	0	50	50
P3	100	50	50
P4	100	0	50
P5	100	0	0
P6	0	50	50
P7	100	100	100
P8	0	100	100
P9	100	100	100
P10	100	0	100
Mean	**70**	**50**	**65**

- number of errors made performing specific tasks;
- number of requests for assistance accomplishing task;
- accuracy of completed tasks;
- objective measure of quantity of output;
- percentage of customers who can carry out key tasks with no training.

How might we collect these data? Take as an example the completion rate: the percentage of customers who successfully complete the task. We could collect these data in a usability test where we ask participants to complete a series of tasks for us.

For example, imagine we have an e-commerce site that sells music CDs. Typical tasks might include:

- Find out if the site carries a compilation CD of Frank Sinatra's greatest hits.
- Find out how much it will cost to buy (including delivery).
- Buy the CD and have it sent directly to a friend as a present.

Assume that we carry out this test with ten typical customers of the site. We might get the kind of data listed in Table 9.1.

The table shows the completion scores for each participant on each of the tasks. We assign a score of 100 per cent if a participant is successful, and 0 per cent if the participant fails. Sometimes, participants get close to succeeding and under this situation we assign a score of 50 per cent. (If you feel uncomfortable with the level of judgement being made here, don't worry about assigning partial completion scores.)

For example, considering the 'Search' task, we can see that participants 2, 6 and 8 (P2, P6 and P8 in the table) were unsuccessful on the task: their completion score was 0 per cent. All the other participants completed the 'Search' task successfully. We can then compute an overall completion score for the 'Search' task: 70 per cent in our example (this is simply the arithmetic mean).

For the 'Price' task, some of the participants scored 50 per cent. This means that they got about halfway through the task – for example, they managed to find out the price of the CD but they did not manage to find out the delivery costs.

We can also compute an 'Overall' task completion score, by averaging the data across all participants and all tasks. For the data in Table 9.1, the overall task completion score is 62 per cent. We can express this in words by saying that any customer has a 62 per cent chance of completing all the typical tasks with our web site. This also means that 38 per cent of customers will probably fail on at least one task, so we have clearly got some work to do on this site.

Measures of efficiency

A very efficient user interface has a high percentage of successful customers in a small amount of time. The reason we consider success and speed together is because people make speed/error trade-offs when using a web site: the faster people work, the more likely they are to make errors; and the more conscientiously people work (to reduce errors) the more likely they are to work slowly. Measures of efficiency allow you to compare fast interfaces that might be prone to errors (e.g. command lines with wildcards to move or copy files) with slower interfaces where errors are less likely or at least easy to spot and correct (e.g. using a mouse and keyboard to drag each file to a different window). The most common measure of efficiency is 'Time on task', but there are other measures such as:

- time to execute a particular set of instructions;
- time taken on first attempt;
- time to perform a particular task after a specified period of time away from the web site;
- cost to perform task (for a business-to-business site, relative to customer's weighted salary);
- time to perform task compared to an expert;
- time to learn task to a particular criterion;
- time to achieve expert performance;
- number of mouse clicks or key presses taken to achieve task;
- time spent on correcting errors;
- number of icons remembered after task completion;
- percentage of time spent using on-line help;
- time spent relearning functions.

Using the same example as above, our 'time on task' table might look like Table 9.2.

This shows for example that the first participant (P1) took 485 seconds to complete the first task ('Search').

You will notice that some of the cells in the table are empty. These cells correspond to participants who did not manage to complete the task (so a

Table 9.2 'Time on task' data (in seconds)

Task	*Search*	*Price*	*Buy*
P1	485	247	647
P2		151	695
P3	359	88	734
P4	426	759	
P5	592		
P6		108	697
P7	332	212	626
P8		288	790
P9	289	104	896
P10	375	628	
Mean	**408**	**171**	**719**
Standard error of mean	**38.9**	**29.6**	**29.2**

'time on task' measure for these participants would be fairly meaningless). You will also notice that the table provides the mean time for each task and also the standard error of the mean (SE), a measure of variability. The standard error of the mean is computed by working out the standard deviation of the sample and then dividing this value by the square root of the number of participants (in our example, the square root of 7 for the 'Search' and 'Buy' tasks and the square root of 9 for the 'Price' task).

What do these numbers mean? For example, if you look at the times taken to complete the 'Search' task, one participant (P7) completed the task in 332 seconds whereas another completed the task in 592 seconds (P5). The average value for the seven participants who completed this task is 408 seconds. These numbers vary quite a lot, and we only tested ten participants. So how would a *typical* customer perform?

The SE is a useful statistic because it helps you make a judgement on the 'true' mean value. Statisticians have shown how to calculate, with 95 per cent probability, that your sample mean is representative of the population as a whole. They have shown that the typical mean value could be as low as the mean minus 1.96 times the SE, or as high as the mean plus 1.96 times the SE. For the 'Search' task, this means the typical mean value lies between $408 - (1.96 \times 38.9)$ and $408 + (1.96 \times 38.9)$, that is between 331.8 seconds and 484.2 seconds. So we can say that a typical customer who manages to complete this task will probably take at least five and a half minutes and not much more than eight minutes.

Dividing the completion rate by the time on task provides us with the core measure of efficiency. This value specifies the percentage goal achievement for every unit of time. For example, for the 'Search' task above, the average completion score was 70 per cent and the average time on this task was 408 seconds (or 6.8 minutes). This yields an efficiency score of 10.3 per cent: this means that for every minute the participant spends on the 'Search' task, just over 10 per cent of the task gets completed.

Box 9.1 Usability testing and statistics

This is probably a good point to distinguish usability testing from experimental psychology. In experimental psychology, statistics are used to make sure you do not draw the wrong conclusion. For example, a result is 'significant' if it is unlikely to happen by chance (chance is usually defined as an event that happens more than five times in 100). Nielsen has pointed out that this rigorous level of decision making is too strict for making usability judgements. He writes:

> Consider the problem of choosing between two alternative interface designs. If no information is available, you might as well choose by tossing a coin, and you will have a 50% probability of choosing the best interface. If a small amount of user testing has been done, you may find that interface A is better than interface B at the 20 per cent level of significance. Even though 20% is considered 'not significant,' your tests have actually improved your chance of choosing the best interface from 50/50 to 4-to-1, meaning that you would be foolish not to take the data into account when choosing. Furthermore, even though there remains a 20% probability that interface A is not better than interface B, it is very unlikely that it would be much worse than interface B. Most of the 20% accounts for cases where the two interfaces are equal or where B is slightly better than A, meaning that it would almost never be a really bad decision to choose interface A. In other words, even tests that are not statistically significant are well worth doing since they will improve the quality of decisions substantially.
>
> Nielsen (1994)

The point is that we are using the data to make practical judgements on whether to proceed with a particular design, not to prove a scientific point.

Measures of satisfaction

The third part of the usability picture is satisfaction ratings. Customers may complete the task and work quickly but they may still dislike the web site.

Measures of satisfaction are certainly 'softer' than the other measures described above. To generate real numbers, many practitioners tend to use a rating scale that is given to the participant after the task. The rating scale might include statements such as the following:

- I had major problems with this task.
- People with a background like mine would be able to carry out this task without much trouble.

0	1	2	3	4	5
Not Applicable	Strongly Disagree	Disagree	Neutral	Agree	Strongly Agree

Figure 9.3 Question response options

- When I looked at the screen it was easy to figure out what to do.
- I often felt lost when using this web site.

Participants are asked to respond to each question by circling one of the answers shown in Figure 9.3.

There are important sources of bias that need to be considered when constructing this kind of questionnaire. In particular:

- Balance the number of negatively-phrased questions and positively-phrased questions (some people prefer to 'agree' or acquiesce in their responses).
- Be aware that early questions may affect the participant's response to later questions (people try to act consistently).
- Keep questionnaires short: if you ask too many questions, participants will lose concentration and not truly think about their answer.

To avoid these sources of bias, follow best practice:

- Pilot test the questionnaire to make sure there are no ambiguous or complex questions.
- When analysing the data, the modal score (the most frequently occurring value) is a better measure of central tendency than the mean.

Other examples of satisfaction measures include:

- ratio of positive to negative adjectives used to describe the web site;
- percentage of customers that rate the web site as 'more satisfying' than a previous version of the site;
- percentage of customers that rate the site as 'easier to use' than a key competitor;
- percentage of customers who feel 'in control' of the web site;
- customer rating on a seven-point scale anchored with 'makes me more/ less productive';
- percentage of customers who would recommend it to a friend after two hours' use;
- customer rating of quality of output;
- rate of voluntary use.

We can now summarise this in the key performance indicator form as shown in Figure 9.4.

Measurement technique	
Scale	Overall task completion score (%) for all tasks in the test suite
Method	Usability performance test with 12 customers at version 0.2 (the first electronic prototype)

Figure 9.4 Key performance indicator: measurement technique for an on-line bank

Setting performance criteria

Our final step in the process is to put values on the scale that we have defined (in the example above, the percentage task completion score). Try to assemble a cross-functional team to agree these values. 'Cross-functional' means that you should aim to get participants from marketing, development, design and project management (at least).

The first trap to avoid is setting a value of 100 per cent for the task completion score. Unless the application is safety critical, this goal is simply not worth attaining. The amount of design effort needed to make sure that all customers will be able to complete all tasks successfully will mean that your site may never see the light of day.

So given that we will not aim for 100 per cent, what value should we put on the criterion?

Making a decision on what value is acceptable requires some market analysis. Useful questions to ask at this stage include:

- Is there an earlier version of the site? If so, how does it score?
- If the task is currently performed without a computer system (for example, by telephone via a call centre), how does that system score?
- How do competitor sites perform?

These values will provide you with a lower limit to put on the performance of your site. Usability specialists then recommend that you consider the response range to be a continuum, ranging from 'Unacceptable', through a 'Minimum' range, into a 'Target' range and then finally into an 'Exceeds' range. This is shown schematically in Figure 9.5.

Figure 9.5 shows how we might set the range of data for our on-line banking example. The example shows four ranges:

- Unacceptable: if the site performs within this range it cannot be released. If it is released it will damage the company's brand and we will lose customers to the competition.
- Minimum: if the site performs within this range, it is barely acceptable. Management must weigh the benefits of releasing a barely adequate site now versus waiting for the usability defects to be fixed.

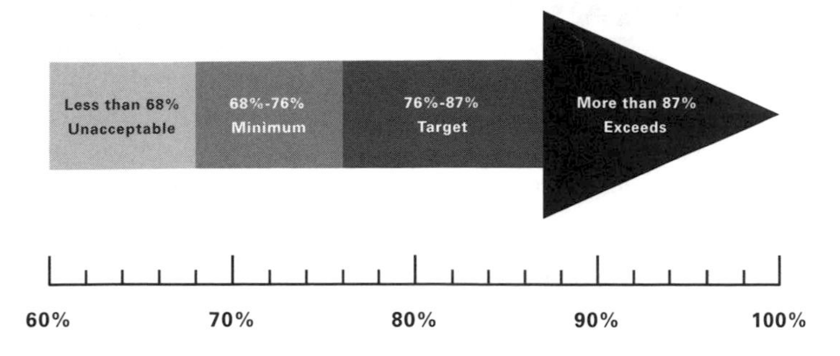

Figure 9.5 Overall task completion score. The values used in this figure are for demonstration only.

- Target: if the site performs within this range, it can be released. This is the performance range needed to succeed in the market place.
- Exceeds: if the site performs within this range, it suggests we have put too much effort into the design of the site (or we have chosen a trivial task). We are probably behind schedule and should release the site as soon as practicable.

Note that this approach ensures that we neither under-engineer nor over-engineer the web site.

Finally, we enter these values into the key performance indicator form (see Figure 9.6).

In practice, you will have a range of key performance indicators for the web site. In the example above, we focused only on task completion: for your site, it may make sense to also measure time on task and satisfaction ratings. For the context of use, we focused only on experienced web users: you may need to consider novice customers of the site too. Finally, there may also be other high-level objectives around which you will want to set key performance indicators.

Number of criteria

How many key performance indicators should a typical project have? The precise number depends on the scope of the project: small projects can be

Performance values			
Unacceptable range	Minimum range	Target range	Exceeds range
Less than 68%	68% - 76%	76% - 87%	More than 87%

Figure 9.6 Key performance indicator: performance values for an on-line bank

successful with just one or two, whereas larger projects will of necessity have more aspects of the site to track. As a rule of thumb, you should have no more than three or four high-level objectives and no more than around ten key performance indicators. Not all of the key performance indicators need to be measured with a usability test (some can be tracked during the project using usability inspection methods).

How to present the information

Appendix 8 provides a form structured to help you specify key performance indicators for your own site.

Summary

- Usability can be measured just like any other engineering attribute.
- Key performance indicators are quality metrics that provide the mechanism for specifying usability values.
- A key performance indicator has four components: a high-level objective, the context of use, a measurement method and performance values.
- The high-level objective is an aspiration statement. It answers this question: 'What needs to happen for the site to be considered a success?'
- The context of use grounds the objective in a real-world task: it says who will be doing the task and under what circumstances.
- The measurement method includes measures of effectiveness (such as task completion), efficiency (such as time on task) and satisfaction (such as an opinion questionnaire).
- Performance values are derived from an earlier version of the site or from competitor sites. The precise values chosen ensure that the web site is neither under- nor over-engineered.

Reference

Nielsen, J. (1994) 'Guerrilla HCI: using discount usability engineering to penetrate the intimidation barrier'. In Bias, R.G. and Mayhew, D.J. (eds) *Cost-Justifying Usability*. Boston, MA: Academic Press.

Further reading and weblinks

Books and articles

Johnson, J. (2000) 'Responsiveness bloopers'. Chapter 7 of *GUI Bloopers*. London: Morgan Kaufman Publishers.

Whiteside, J., Bennett, J. and Holtzblatt, K. (1988) 'Usability engineering: our experience and evolution'. In Helander, M. (ed.) *Handbook of Human-Computer Interaction*. Oxford: North-Holland.

Web pages

McMillan, B. (undated) 'Usability specifications'. http://ijgj229.infj.ulst.ac.uk/BillsWeb/HCI/Lectures/lect9.html. Lecture notes describing how to set usability goals for a system.

Nielsen, J. (2001) 'Usability metrics'. http://www.useit.com/alertbox/20010121.html. 'Although measuring usability can cost four times as much as conducting qualitative studies (which often generate better insight), metrics are sometimes worth the expense. Among other things, metrics can help managers track design progress and support decisions about when to release a product.'

Perfetti, C. (2001) 'The truth about download time'. http://world.std.com/~uieweb/truth.htm. 'We hear all the time from web designers that they spend countless hours and resources trying to speed up their web pages' download time because they believe that people are turned off by slow-loading pages. What we discovered may surprise you.'

10 Develop the information architecture

Why architecture?

'Halfway through the book and we can finally start on *design*.'

It is very difficult to begin a web project and prevent yourself from sketching out potential solutions on a blank sheet of paper. We are now almost halfway through the book and we haven't yet put pencil to paper (or opened PhotoShop). This is the stage when 'design', as most people think of it, actually happens.

However, you will have to wait before we start talking about colours, fonts, icons and screen layouts. That is the substance of the next chapter. In this chapter we will discuss the structure of the information behind the web site and how we intend to navigate through it.

The term we will use for this is 'information architecture'. At first, this may seem a rather grand, even pretentious, phrase. But there are a number of reasons why the term is useful.

Imagine that you are an architect about to design a building for a client. Your initial questions will address the location of the building. Where will it be? Are there good transportation routes (such as roads and public transport) so that people can get to the building?

Next, you will want to know about the size of the building. To answer this question, you will need to know its purpose: a company that makes products will probably need large empty spaces to provide warehousing facilities, whereas a company that sells insurance will need smaller, self-contained offices. How many floors? How many rooms on each floor? If you make the wrong decisions at this point, the building may very quickly feel too small for its purpose. Once it is built, you might need to retrofit extra office space.

You might also want to consider how people move from floor to floor and from room to room. If there is an elevator, there will need to be a lobby on each floor where people can enter and leave the elevator. You will also want to include stairs so that people can leave the building in an emergency.

You will also want to think about the generic 'systems' that are common to each floor and each room of the building: electricity, telephone and computer networks and water pipes for example.

This serves as a useful analogy for building a web site. For location, read domain name (is it memorable? Can people find it?) and bandwidth (will it support the number of visitors you expect to have)? For size, read number of pages and the way these are linked together. For movement within the building, read navigation within the web site. For generic systems, read 'Search', 'Copyright information' and 'Help'.

Information architecture – the structure of your site – needs to be clearly distinguished from screen layouts – the individual pages that the customer sees on screen. The architecture issues need to be resolved before the detailed screen design can take place – or you may find yourself with an information architecture that cannot evolve to include your next range of products, your new ways of working, or the new delivery channels you decide to introduce in a few years' time. Retrofitting these features will be expensive and will almost certainly confuse returning customers who have got used to your earlier site.

In this chapter, we will apply the principles to a specific application domain. We will design an e-commerce site that aims to sell downloadable music. The music is in electronic format and will be delivered to customers via the Internet (so customers will not receive a CD through the post). For the moment, suspend disbelief and assume that all of the copyright and copy protection issues have been resolved. The opportunity has been analysed (it is very favourable) and the context of use has been defined. Key performance indicators have been agreed.

Derive a content inventory

The first step of the process is to explicitly list the likely content of the site from the perspective of the customers' tasks. We will look at two types of content, critical content and secondary content.

Identify critical content

In this step, you refer back to the task list from the context of use data collected earlier. The aim is to identify the critical tasks that your music-buying personas engage in when purchasing a CD.

These might include:

- browse all the available music within a genre;
- look for a particular artist;
- compare similar music on different CDs (for example, the customer who likes classical music might compare CDs of a specific classical work to help choose between different orchestras or conductors);
- read liner notes and track listings;
- find out the price of the music;
- buy the music.

This provides us with specific ideas about the main content that we will need to include in the site. For example, we will need a list of artists and their music, prices for each of the items and an e-commerce engine to support the transaction process.

Identify secondary content

We now identify the secondary content that our context of use analysis has identified as important to our specific personas. The ideas for this secondary content will have come from interviews with potential customers, a review of competitor sites and ideas from the design and marketing teams. For our site, it might include:

- read music reviews;
- listen to music samples;
- read interviews with artists;
- chat with people who like the same music;
- read potted biographies of artists;
- browse a gig guide;
- register for e-mail updates of new releases;
- see the fantasy record collections of music gurus;
- have an artist or album recommended (based on listening preferences);
- see this week's top ten downloads;
- see the members' all-time 'Top 100';
- buy tickets for a concert;
- read a 'clubber's guide';
- buy a book about an artist;
- read copyright details.

The actual list may be very long – containing 30 or more headline tasks. None of these are critical to the purchase of music so we identify them as secondary. (Indeed some of the content, such as copyright information, may be irrelevant to most of our customers but we may need to include it for legal reasons.)

Categorise the information

In this step, we try to work out how customers categorise the content that we want to present on the site. This will provide us with the gross layout of the site and help us begin to specify an initial navigation framework.

There are a number of ways to do this, but probably the simplest is to use a card sorting technique. Take a set of index cards or a pack of sticky notes and separately on each one write down an item of primary or secondary content. Then ask some customers to group the information together. For example, with the content listed above a partial set of results might look like Figure 10.1.

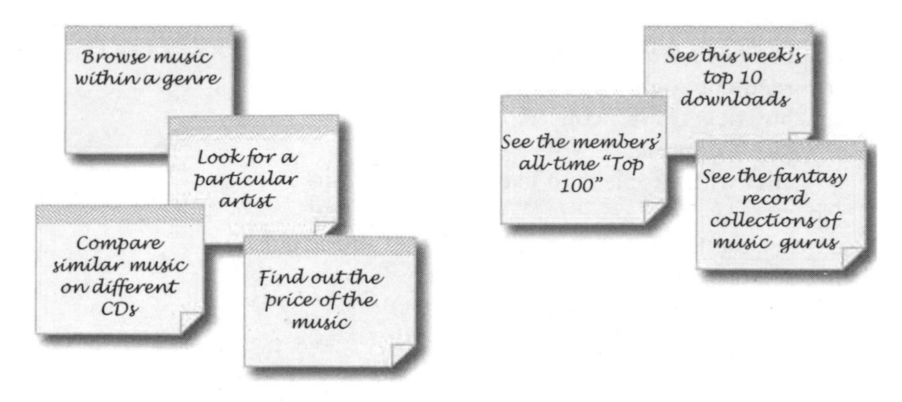

Figure 10.1 Card sorting technique

You can see from Figure 10.1 that certain information has been grouped together; for example, the right hand group contains the following cluster of notes:

- See the fantasy record collections of music gurus
- See this week's top ten downloads
- See the members' all-time 'Top 100'.

When the customer has finished, ask for a term or title to describe the cluster. For example, the term given to the cluster above might be 'Listings or charts'. This term provides useful input when deciding on the precise terms to use for items in the user interface. Note that these sticky notes do not represent individual pages in the final design – the final design may need a number of screens to support a task (for example, it is unlikely that the 'Browse music within a genre' task could be supported by just one screen).

Let us assume that, for the list above, cluster analysis revealed the following structure:

1 Browse
   a browse all the available music within a genre;
   b look for a particular artist;
   c compare similar music on different CDs;
   d view track listings;
   e read liner notes;
   f find out the price of the music.
2 Research
   a have an artist or album recommended (based on listening preferences);
   b listen to music samples;
   c read potted biographies of artists;
   d read interviews with artists;
   e read music reviews;
   f buy a book about an artist.

3 Interact
 a register for e-mail updates of new releases;
 b read copyright details;
 c buy tickets for a concert;
 d buy the music;
 e chat with people who like the same music.
4 Charts
 a see the fantasy record collections of music gurus;
 b see this week's top ten downloads;
 c see the members' all-time 'Top 100';
 d read a 'clubber's guide';
 e browse a gig guide.

At this point, we also need to identify other key screens that we will need to support the required functionality, for example: 'Search' and 'On-line help'.

Note at this stage that we are simply organising the functionality into categories. Navigation between the categories is the purpose of the next step.

Define the navigational framework

The content inventory, derived from the context of use information, provides us with a good idea of the overall functionality for the site. We now need to define the pathways between this information. We need to decide how people will move between the major areas of the site ('global navigation') and how they will move within each of these areas ('local navigation').

For example, a store selling books, CDs and movies might define the major areas of the site as 'Books', 'Music' and 'Video'. Navigation within these areas (for example, within 'Books') would be defined by a local navigation scheme that might include 'Book search', 'Browse categories', 'Bestsellers' and 'Paperbacks'. Within this scheme there may be an additional level of navigation: for example, 'Browse categories' would include a navigation scheme that includes 'Biography', 'Fiction' and 'Humour'.

Global navigation

Global navigation provides a framework for moving between the major areas of the site. We can turn to the results from the previous step for this information, and in particular look at the high level categories that we identified. In our example, we have four of these categories, 'Browse', 'Research', 'Interact' and 'Charts'. This provides us with the global navigation items.

We also noted that we need to include 'Search' and 'On-line help'. These are distinct areas of functionality and we make a note that we might want to present these in a different way to the other task-oriented items. If we consider this visually, our structure looks like Figure 10.2.

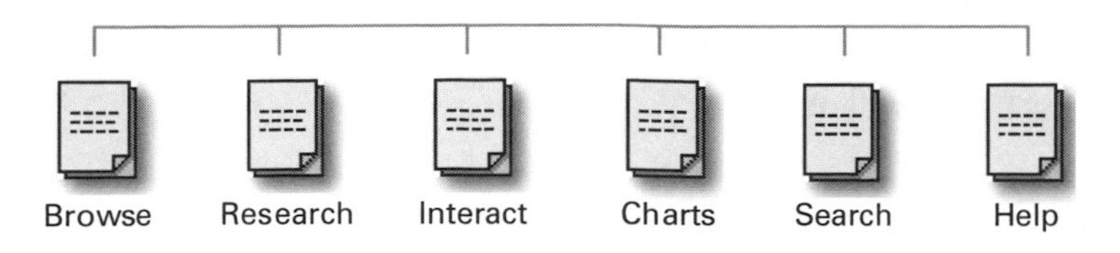

Figure 10.2 Global navigation

This shows us the main categories or links that we should provide on every page of the site.

Local navigation

We now look at the navigation within an area. Take the 'Browse' section of our site as an example. We defined the main areas of this site as supporting the following tasks:

- browse all the available music within a genre;
- look for a particular artist;
- compare similar music on different CDs;
- view track listings;
- read liner notes;
- find out the price of the music.

At this point it is worth sketching out a potential navigation path (as in Figure 10.3).

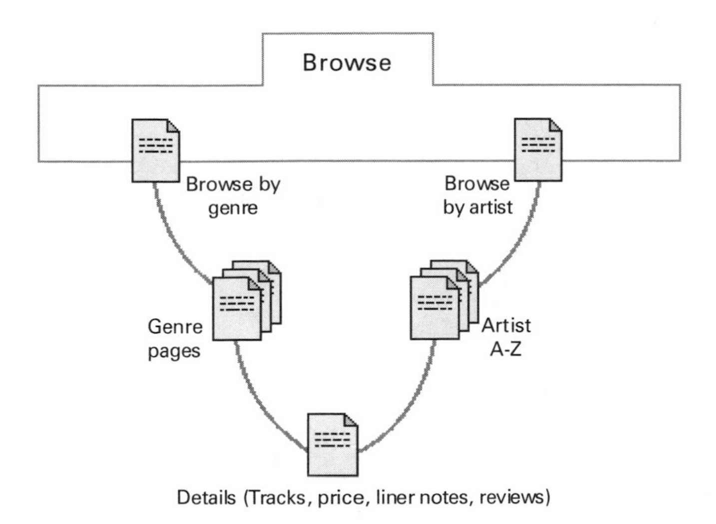

Figure 10.3 Local navigation

Note that this is just one solution to the problem. Later we will get the opportunity to test out our assumptions about the structure of the site with customers, and it will almost certainly change before coding starts.

Develop high-level functional mock-ups

In this step, we take the key screens and sketch out some potential solutions. These mock-ups have little content: we use them to check the site navigation.

At this point it is important to ignore the detailed, visual design issues. This is not the time to think about branding the site. We simply sketch out three or four different high-level designs to support the navigation framework developed in the previous steps. These designs will look rough but that is fine. In this step, we might try to answer questions such as these:

- Will the products be in a hierarchy that customers browse, or are they in a large (flat) catalogue that customers search?
- Will the site provide a way to compare items with each other, feature by feature?
- Will the site make recommended purchases based on the customer's buying habits? Or based on preferences volunteered by the customer?
- Will the site provide user accounts – and will customers have to sign in?
- Will people need to save information across sessions (such as lists of items they are considering buying)? If so, will we support this by using a shopping cart metaphor (without requiring a purchase), or do we need to add a new feature (such as 'My favourites')?

In this step it is important to ensure that the design matches the customer's 'mental model'. This means that the design should mirror the customer's thought processes in the way it organises information and structures the customer's tasks.

Decide the user interface metaphor

User interface metaphors are so common in personal computing that we sometimes forget that they are metaphors. The dictionary defines a metaphor as 'The application of a name or a descriptive term or phrase to an object or action where it is not literally applicable'. Within the domain of computing, perhaps the most prevalent metaphor is the 'desktop metaphor' used on Macintosh and Windows operating systems.

Metaphors are abstractions that help customers predict how something works. The point of the desktop metaphor is to leverage the user's knowledge of files, folders, and trash cans to help them manage the objects on their computer: for example, files are put in folders, or in the trash when they are no longer needed.

Web sites use metaphors too. Examples of metaphors used on the web to support navigation include:

- hypertext metaphor;
- index card metaphor;
- computer metaphor;
- visual metaphor.

Hypertext metaphor

The hypertext metaphor is the very basis of the web. The metaphor simply states that clicking on an underlined word or phrase will take you to a new page. All sites on the web use hypertext for navigation, although some use it less extensively than others. Yahoo! (www.yahoo.com) uses the hypertext metaphor extensively for global and local navigation.

This type of navigational metaphor has a number of strengths. First, it will work in any browser. Second, there is no ambiguity over what is a link and what is not a link (although when a link spills over onto a new line bullets need to be used to make line breaks unambiguous). Third, anyone that can use the web will know how to navigate this site. Finally, it is accessible by people with disabilities.

This navigation metaphor is effective and efficient. However, it can appear plain and unexciting and some people argue that this may affect the customer experience. Other sites have therefore chosen to alter the metaphor, such as the BBCi (www.bbc.co.uk) site. Hypertext links are neither underlined nor in blue. Although aesthetically more pleasing, this metaphor assumes that customers will take the time to distinguish links from non-links. Inevitably some customers will waste time clicking on words or phrases that are not links, or 'minesweeping' to find precisely what is clickable.

Index card metaphor

A second common navigation metaphor on the web is the tab-based metaphor. The tab metaphor is based on the use of index cards in (for example) a recipe box. By clicking on one of the tabs it comes to the 'front', in the same way that selecting 'Desserts' in a box of recipe cards brings that card to the front. This type of navigation bar also makes it clear where you are in the site (for example, 'Home'). Amazon (www.amazon.com) is a good example of the use of the tab-based metaphor.

Tabs are a useful metaphor for organising information, but if you use this approach be careful not to use too many tabs. In particular, try to avoid multiple rows of tabs (where one row of tabs is tacked above another) since this destroys the very metaphor on which it is based.

Computer metaphor

A third navigation metaphor common on the web is more solidly computer based. Sites that use this metaphor use menus or buttons to aid navigation.

Screens look and behave very much like a Windows application. Good examples of this can be seen at the *National Geographic* site (www.nationalgeographic.com).

Visual metaphor

The other navigation metaphor used on the web is a visual metaphor. Examples of this include the OXO site (www.oxo.com) where the navigation frame is itself a noughts-and-crosses grid. Other examples include the Jimtown store (www.jimtown.com), which uses a two-dimensional plan view of the store within which customers navigate; and taobot (www.taobot.com), which uses a three-dimensional building as its navigation metaphor.

Visual metaphors can be problematic. For example, the layout of the real Jimtown store may have been made because of physical or practical constraints. Certain decisions, such as putting the deli near the kitchen, were not made to make it easier for shoppers to find their way around the store – the decision was probably made for practical reasons. Reproducing limitations from the physical world in the on-line version reveals a significant limitation of visual metaphors.

Examples

Using these metaphors, three possible examples of the 'Browse' page are shown in Figures 10.4, 10.5 and 10.6.

Figure 10.4 shows a solution that uses a file manager metaphor within a tab-based structure. The genres (for example, Blues, Classical, Country etc.) are listed in the first column; selecting a genre (for example, 'Blues') lists the sub-genres (such as 'Bestsellers'); selecting a sub-genre lists the music itself. Selecting a title might then load a new page showing more details about that piece of music (such as the track listings).

In contrast, the high-level design shown in Figure 10.5 uses a hierarchical, 'Yahoo!' type metaphor. Each of the genres and sub-genres are explicitly listed; selecting one of these takes the customer to a new page that lists the music within that sub-genre. Note also that the navigation convention is different on this page: rather than the tab metaphor used in Figure 10.4, the interface uses a set of buttons for the four key areas of the site.

Figure 10.6 shows a third design that combines parts of Figures 10.4 and 10.5. One difference is that this design has a frame on the left-hand side that remains persistently visible. This contains icons for each of the high-level navigation items. This has allowed more space on the page for additional genres. (Note that in both Figures 10.5 and 10.6, fake Latin represents missing text – this simply says that text will be placed here, but we have not yet decided what that text will be.)

A fourth metaphor that we have not explored here is a visual metaphor, for example a record store. We could sketch out a design that represents a

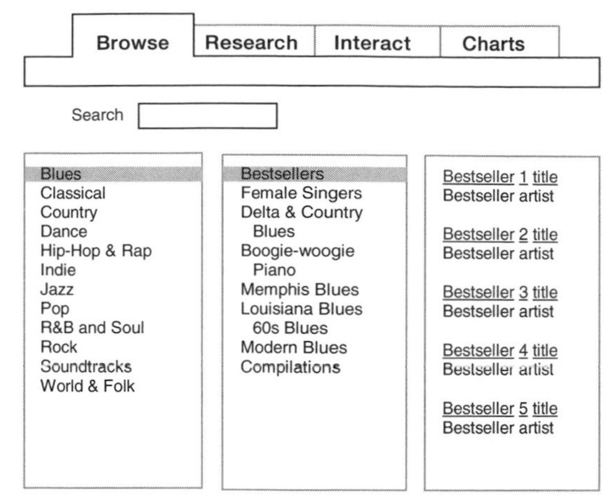

Figure 10.4 Browse page based on a file manager type metaphor

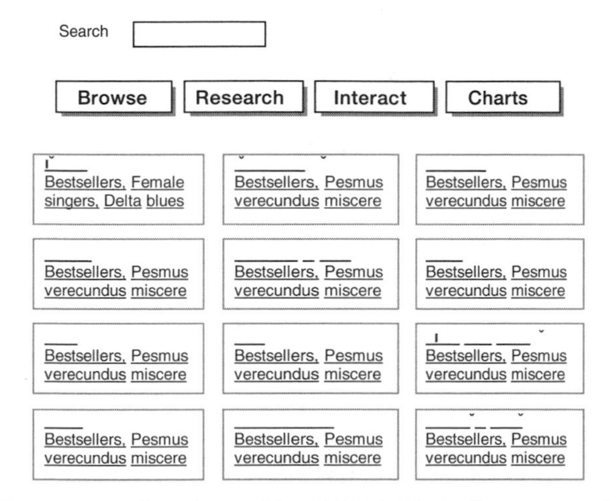

Figure 10.5 Browse page based on a hierarchical 'Yahoo!' type metaphor

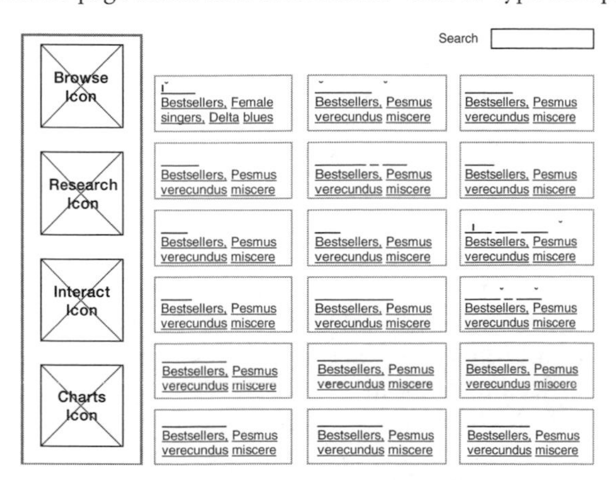

Figure 10.6 Browse page combining file manager and hierarchical metaphors

plan view of a record store, with separate visual areas for 'Browse', 'Research', 'Interact' and 'Charts'.

Test against task scenarios

Now we have some high-level prototypes and we need to choose between them. The best way to do this is to show the designs to a small number of customers and ask them to carry out one of the tasks defined in the context of use. For example, you might ask a customer to buy the Top 10 Blues downloads.

How should you implement the prototype? You may be tempted to develop an electronic prototype at this stage. However, it is just as effective (and certainly much quicker) to produce sketches on paper.

One criticism often levelled against paper prototypes is that they look 'amateurish'; designers do not like showing them to customers in case they think badly about the company or site. In practice, customers quickly learn to 'play the game' and get fully involved with the prototype. Once a 'button' has been pressed you present the next sketched screen and so on. You do not need to sketch out a new screen for every stage of the interaction: for example, sticky notes can be used to simulate pull-down menus.

Moreover, because the prototype is on paper, customers realise that every aspect of the design is open to change. This makes them more willing to criticise a part of the design that they find unwieldy. In contrast, when using an electronic prototype, customers may think they are seeing the result of many days' coding and development; in this situation, some of them will be unwilling to complain too loudly. (Chapter 12 describes some testing methods in more detail.)

After testing your prototypes, you will usually find that one performed particularly well. The other prototypes probably had one or more features that also showed promise. You now need to revise your prototype in the light of these comments, and then we can lay out the screens. This is the topic of the next chapter.

How to present the information

Once you have converged on an agreed information architecture, this high-level design needs to be documented. The level of detail depends on the size and scope of the project. For small projects where the design team is co-located, you could mock up an interactive, wireframe prototype (for example, in applications such as Microsoft PowerPoint or Macromedia Director). Or you could simply scan in your final sketches and describe the behaviour of each screen in a short report. For larger projects (or for particularly complex screens) you may need to do both.

Summary

- The information architecture describes the high-level, functional layout of the web site.
- We start by deriving a content inventory: this lists all the types of information that you will have on the site.
- Next, we categorise the information by finding out how customers organise the information in their heads.
- We then use this information to define a navigational framework for the site: the pathways and links between the various content areas.
- This information is then assembled into high-level, functional mock-ups: these are prototypes, sketched out on paper, showing alternative ways of presenting the information.
- These prototypes are then tested with customers.

Further reading and weblinks

Books and articles

Reiss, E.L. (2000) *Practical Information Architecture: A Hands-On Approach to Structuring Successful Web Sites*. Harlow: Addison-Wesley.

Rosenfeld, L. and Morville, P. (1998) *Information Architecture for the World Wide Web*. Sebastopol: O'Reilly & Associates.

Shedroff, N. (2001) *Experience Design 1*. Indianapolis, IN: New Riders Publishing.

Web pages

Nielsen, J. (2000) 'Is navigation useful?'. http://www.useit.com/alertbox/20000109.html. 'For almost seven years, my studies have shown the same user behavior: users look straight at the content and ignore the navigation areas when they scan a new page.'

Toub, S. (2000) 'Evaluating information architecture'. http://argus-acia.com/white_papers/evaluating_ia.html. 'This white paper explores the why's, what's, and how's of evaluating a web site's information architecture.'

11 Lay out the screens

Introduction

We now have a functional prototype of the site and need to organise the information on screen. This chapter contains guidelines for the process of transforming the prototype into screens, complete with data entry and display fields. This chapter focuses specifically on the visual design of pages that aim to support e-commerce transactions.

The process comprises four principal stages:

- transform the functional objects into screen objects;
- select labels for each screen object;
- create additional screen items;
- arrange objects and labels on the screen.

These stages are sequential but, as with all design activities described in this book, should be part of an iterative development process. An iterative process is essential to ensure a suitable balance among the number of screens, their individual contents and the flows within and between screens.

Transform the functional objects into screen objects

This section is organised around a flow chart (Figure 11.1) illustrating key design choices concerning the design of items for the screen. Within the flow chart are a number of points at which design questions must be asked. Along the flow there are a number of end-points where appropriate screen objects are recommended. These screen objects are those commonly available to the web site designer.

Within standard HTML there are a defined set of standard controls. These are:

- radio buttons;
- check boxes;
- drop-down/pop-up list box;
- single/multiple selection list box;
- single/multiple line text box.

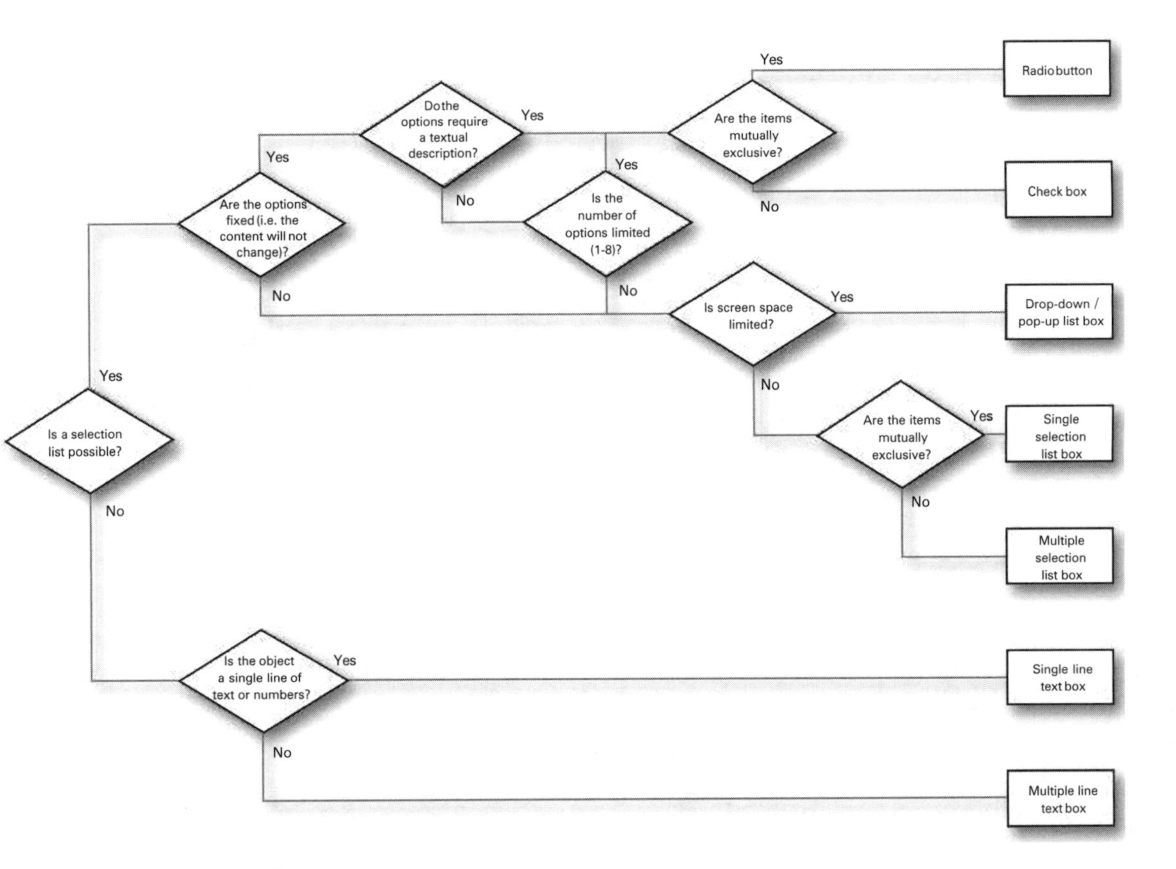

Figure 11.1 Screen item selection flow chart

This is a narrower selection than the designer might ordinarily have when designing screens for other application environments, such as Microsoft Windows. For example, it excludes the 'spin box', a control that allows the user to increase or decrease the value in a single line field by clicking on an up or down arrow. Programming languages such as Javascript and browser plug-ins such as Macromedia Flash make this, and other, controls available, but realistically the standard HTML selection is adequate for any user interface.

Yet despite the smaller selection of controls, there are many examples on the web where these controls are misused. The aim of the flow chart is to help you pick the right control.

For example, imagine that you are designing a questionnaire and you want to ask customers which browser they usually use (for example, Internet Explorer, Netscape Navigator or some other browser). Interfaces provide you with a range of controls that you can use to help customers enter data. The guiding principle is to design a data entry system that customers can use quickly and that minimises errors. The flow chart in Figure 11.1 will help you achieve this.

The flow chart shows that the first design question we ask is, 'Is a selection list possible?' Since the number of possible names is short (three in our example), we choose Yes. Next, we ask: 'Are the options fixed?' Since the content is pre-defined, we answer Yes. The next question we answer is: 'Do the options require a textual description?' Since our data are in the form of text (rather than numbers) we answer Yes. Finally, we ask if the items are mutually exclusive. In our case, since we are interested in the most commonly used browser, we again answer Yes, and learn that the best control to use is a radio button.

The subsequent sections describe each of these controls.

Radio button

What is it?

A set of small circles labelled with choice descriptions. When a choice is selected a small dot appears in the chosen circle and any previous choice (in that set) is de-selected. See Figure 11.2.

Purpose

To select one from a small set of mutually exclusive options.

Advantages

- Provides easy access to and comparison of choices
- Allows a textual description to be added to each choice item.

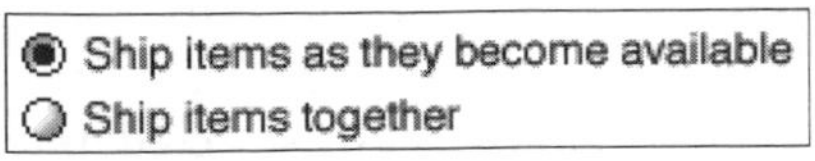

◉ Ship items as they become available
○ Ship items together

Figure 11.2 Radio buttons

Disadvantages

- Only effective for a limited number of choices. It is not recommended for use where the number of choices is greater than eight.

Best practice

- Display field label as a single line of text
- Position each button label to the right of the button
- Include at least one space between the label and the button
- Grey out unavailable choices
- If necessary, include a 'none' choice to add clarity
- Vertically arranged buttons are generally easier to use
- When horizontal arrangements are used, put a minimum of three character spaces between choices (to avoid confusion with adjacent buttons)
- Enclose the group of buttons in a frame
- Show default selection if appropriate.

Check box

What is it?

A square box labelled with a choice description. The check box is either selected or not. A selection will generally signify a positive response (e.g. 'yes', as in yes/no, or 'on' as in on/off) depending on the context. A tick in the box indicates a selection. An empty box represents a non-selection. See Figure 11.3.

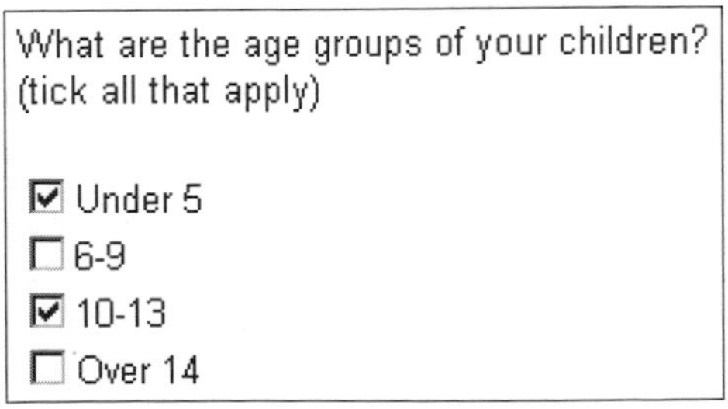

What are the age groups of your children?
(tick all that apply)

☑ Under 5
☐ 6-9
☑ 10-13
☐ Over 14

Figure 11.3 Check boxes

Purpose

To set properties or values where more than one choice can be selected and data choices are small, fixed in number, require a textual description, and are best understood when the alternatives are visible and fixed in content.

Advantages

- Easy access to options
- Easy comparison of alternative options.

Disadvantages

- Consumes screen space
- Can only be used for a limited number of choices
- Single check boxes are difficult to align with other screen controls as they often possess long descriptions.

Best Practice

- Display label in a single line of text
- Position each choice description to the right of the button
- There should be at least one space between the choice description/label and the button
- Signify that a button is unavailable for selection by greying or dimming the label
- Provide meaningful choice descriptions clearly describing the values
- If it is appropriate to show a preset state, display a check/tick in the check box
- Related check boxes should be arranged into groups
- Generally, vertically arranged check boxes are easier to use
- Enclose the group of check boxes in a frame with a clear title
- Arrange the choices/selections in the order expected by the customer or if there is no expected order, arrange by frequency of occurrence or sequence of use.

Drop-down/pop-up list box

What is it?

A single rectangular box with a small button (e.g. with a down-arrow icon) that reveals a larger associated box containing a previously hidden option list when activated. The selected item is displayed in the box after selection. See Figure 11.4.

Figure 11.4 Drop-down/pop-up list box

Purpose

To select one item from a large list of mutually exclusive options when screen space is limited. It is most useful for selecting values, or setting properties, where choices are mutually exclusive or where screen space is limited. This is a good choice of control for data options that are:

- Best represented textually
- Infrequently selected
- Neither well known, easily learned, nor easily remembered
- Ordered in a non-predictable way
- Large in number
- Variable or fixed in list length.

Advantages

- Unlimited number of choices
- Reminds customer of available options
- Conserves screen space.

Disadvantages

- Requires an additional action to display the option list
- May require scrolling in order to access all the list items
- The list may be ordered in an unpredictable way, making it hard to find items.

Best practice

- Left align list options in a column
- Display the descriptions using mixed-case letters
- Present all available options
- Box should be long enough to display six to eight options without scrolling. If more than eight are available, provide vertical scrolling

- Box should be wide enough to display the longest entry
- Order in a logical and meaningful way to permit easy browsing
- Position the label either to the left of the box (preferred) or left-justified above the box
- When the drop-down/pop-up box is first displayed, provide a default choice and display it highlighted
- Use when screen space or layout considerations make radio buttons or fixed list boxes impractical.

Single/multiple selection list box

What is it?

Permanently displayed box-shaped control that contains a list of options from which either single or multiple selections are made (see Figure 11.5).

Purpose

To select from a large set of mutually exclusive or non-mutually exclusive options. The list box may be used for selecting values or setting properties.

Most useful for data and choices that are:

- Best represented textually
- Infrequently selected
- Neither well known, easily learned nor easily remembered
- Ordered in a non-predictable fashion
- Frequently changed (i.e. the list is frequently changed)
- Large in number
- Fixed or variable in list length
- Where the screen space or layout constraints make radio buttons or check boxes inappropriate.

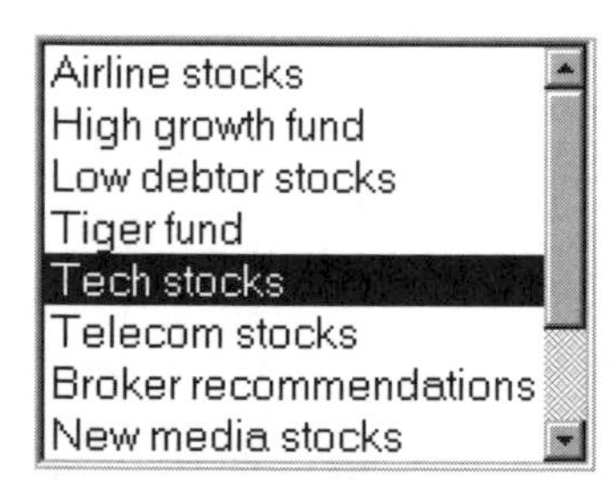

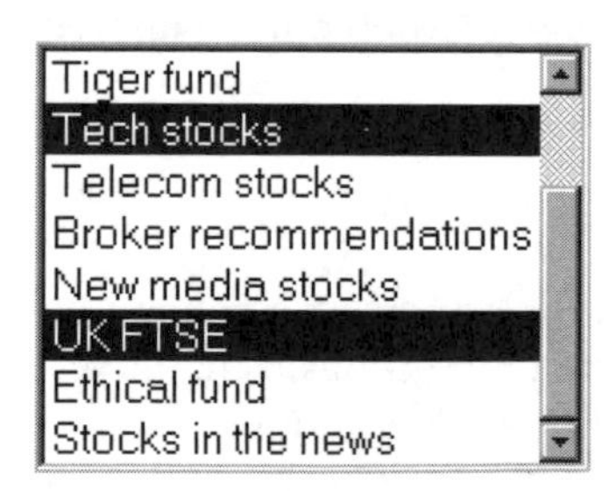

Figure 11.5 Single and multiple selection list boxes

Advantages

- Unlimited number of options
- Customers are aware of the options (recognition, not recall)
- The box is always visible.

Disadvantages

- Scrolling is often required (which is an additional action).

Best practice

- Left-align the list items into a column
- Present all available alternatives
- Box size should be large enough to display six to eight options without requiring scrolling
- If the list is longer than the display, provide vertical scrolling
- The box display should be wide enough to accommodate the longest possible option
- List options according to frequency of use (preferred) or alphabetical order
- Provide command buttons for 'select all' and 'deselect all' actions, when these actions must be performed frequently or quickly.

Single/multiple line text box

What is it?

A text box is a field, often rectangular, in which text (and numbers) may be displayed, entered or edited. Read-only text boxes have a grey background and editable text boxes have a white background. Text boxes may comprise one (single) or more (multiple) lines. An example is shown in Figure 11.6.

Purpose

Permits the display, entry, or editing of textual (alphanumeric) information or the display of read-only information. It is appropriate for data that is not limited in scope, difficult to categorise and varied in length (hence unsuitable for a selection list).

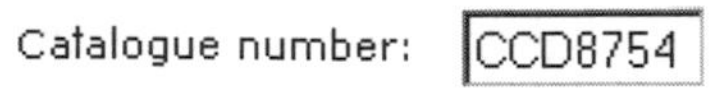

Catalogue number: CCD8754

Figure 11.6 Single line text box

Advantages

- Very flexible
- Familiar to customers (as it is similar to a paper-based entry field)
- Occupies little screen space.

Disadvantages

- The customer must recall the text to enter rather than simply recognise the correct entry
- There is scope for data-entry errors such as misspelling
- Requires the use of a keyboard.

Best practice

- Textual data should be left-justified
- Numerical data should be right-justified
- For single line text boxes, labels should be positioned either to the left of the box or above it and left-justified
- A colon (:) should be positioned immediately after the label. When the label is left of the box, there should be a minimum space equivalent to one character between the colon and the text box.

  For read-only boxes
  - If the field displays alphanumeric data which is short, left-justify the label above the box
  - If the data field is numeric and variable in length, right-justify the label above the box
  - Break up long text boxes through the incorporation of slashes, dashes or spaces (or other common delimiters, e.g. parentheses).

  Field size
  - Text boxes for fixed-length data should be large enough to contain the entire entry
  - Text boxes for variable-length data should be large enough to contain the majority of entries
  - Where entries are potentially larger than the entry field, scrolling should be provided to permit typing into, or viewing of the entire field
  - Utilise 'wrapping' for continuous text in multiple-line text boxes.

Select labels for each screen object

We now have a selection of screen objects and we now need to label each of them.

Choosing the right label

All of your screen objects should be labelled unless their meaning is obvious and can be understood without labels. Care must be taken when selecting labels in order to ensure consistency and to maintain a logical flow between items.

The phrasing of labels should be consistent within each page and across the site. A collection of diverse types of phrasing within a screen will confuse the customer (e.g. mixing statements with questions).

Mark optional fields clearly

If a field is optional, mark it clearly with the word 'Optional' or with another standard marker. When optional fields are grouped with required fields, place the required fields first.

Use intelligent scripting to avoid complex interdependencies ('if/then' rules). For example avoid syntax such as 'If you selected "Yes" in field 1, then fields 2 and 4 should be left blank; if you selected "No" then field 2 should contain ...' etc. Instead, constrain customer choices based on the customer's answers to previous questions.

Provide explanatory messages for fields

Use tool tip or balloon text to further explain the information that the customer should enter, with an example. Do not use tool tip or balloon text to simply repeat the field label. For example, the tool tip text for a 'Date of Birth' field should not simply say 'Enter your date of birth'. Instead write, 'Type your date of birth in the format dd/mm/yyyy, e.g. 18/10/1980'.

Labels: best practice

- Labels should explain the purpose and content of the given object and should be consistent across screens and tasks
- All the labels should be grammatically consistent
- Labels should be clearly distinguishable from the information that they are designating. One way to achieve this is by ensuring a clear space between the label and its object
- Labels should be formatted (e.g. in terms of font and size) and justified consistently
- Units of measurement (e.g. pounds, days, etc.) for the displayed information should be located either within the label or to the right of the field unless the unit of measurement is obvious to the customer
- Aim to avoid the use of jargon (e.g. 'Submit')
- Use standard alphabetic characters. Try to avoid using characters such as '#'
- Use short familiar words – text should use the terminology customers typically use to perform their tasks

- Wherever possible, aim to use labels phrased in positive rather than negative terms
- Arrange all the words in a label on a single line
- Do not hyphenate words if possible
- Do not include punctuation in abbreviations, mnemonics, and acronyms
- Use terms and abbreviations consistently throughout the transaction form. For example, instead of varying terms such as 'National insurance number', 'NI No.' and 'Nat. Ins. Num.', use one term, such as 'NI number'.

Create additional screen items

In addition to screen objects and labels, there are a number of other items that may need to be included in the screen design in order to help the customer interact fully with it. The main additional items are:

- titles;
- command buttons;
- navigational cues;
- prompts;
- additional (or pop-up) windows;
- error messages.

Titles

Titles help the customer navigate through the tasks and to maintain a clear understanding of where they are in the task flow. The screen name should appear in the browser bar (using the HTML <TITLE> tag). The title should be a sensible phrase for bookmarking.

A second type of title is for groupings of objects within the page. These titles should communicate the collective purpose of the items contained within the framed group. They should be located in a position that clearly links them with the group.

All titles should be concise and written in the customer's language, avoiding terms, phrases, abbreviations or acronyms that are not considered familiar to the customer.

Command buttons

Command buttons (see Figure 11.7 for an example) are normally grouped at the bottom of the page, in one or more rows. Have just one row if possible, and no more than three. If the form is long and will require scrolling, duplicate the button that represents the default choice near the top of the screen.

Figure 11.7 Command buttons

Best practice

- Buttons should be clearly separated from other controls and graphics, for example with a horizontal rule or by plenty of space.
- When there are multiple pages to a form, use the same layout for buttons on each one.
- Associate a label with a given action and position on the form consistently throughout the application.
- Use text to label the button, not an icon.

Navigation cues

All pages should provide an 'escape route' back to the home page and the current web convention is to use a company logo in the top left of the screen to achieve this. If the customer is expected to navigate extensively within the site, a 'breadcrumb' trail (a visual cue that shows how this page fits within the navigational framework, see http://keith.instone.org/breadcrumbs for more discussion) may also help. A breadcrumb trail shows the location of a page within the site's hierarchy – 'you are here' – and gives customers a way to go 'up' to higher sections of the site.

Not all web users grasp the idea of hierarchy from a breadcrumb trail, but the concept of progression, or stepping through a series of screens, is important when customers are carrying out a specific task. For example, when completing a multi-page form, show customers where they are in the process. At the very least, provide a textual indicator such as 'Page 2 of 6'.

If the form has more than one page, include 'Back' and 'Next' command buttons (see Figure 11.7) so the customer can return to previous pages (e.g. to check or change previous entries). The customer should be able to move backwards and forwards through the screens without losing input.

Prompts

A prompt is an item of text that provides guidance on performing actions in certain screen areas.

Prompts should be used to inform the customers that there is a need to perform an action (if it isn't obvious to customers that they need to do it), how to perform an action (if the method isn't clear) or what the result of an action will be.

However, prompts should be used sparingly and only in situations where customers are predicted to have problems. Overuse of prompts can make the whole dialogue confusing and cluttered, may distract attention away from important fields/information and is particularly irritating for experienced customers.

Best practice

- Aim for brevity: use short, familiar, and complete words and phrases. Additional instructions can be provided as help screens.
- Use the active voice: for example, 'Type your credit card number:', or 'Credit card number:'; not 'You should type your credit card number'.
- Use simple action words, put the result of the action first and use a mix of upper and lower case letters, e.g. 'To complete your purchase, press the <Place order> button'.
- Try to phrase the action in positive rather than negative terms.
- Use the language of the customer.
- Use standard alphabetic characters.
- Use consistent grammar.

Additional (or pop-up) windows

Additional windows are commonly used for functions that are not part of the customer's main task, such as help and dialogue boxes. Overuse of additional windows can be very confusing, so be sure to restrict their use.

Where possible, help screens should be designed to ensure that there is no need to scroll. In exceptional circumstances where the help contains more information than can be contained on a single screen, scroll bars should be included and the window should also be resizable.

Help screens should not be modal: the customer should be able to use the main window while the help screen is open. It should, however, always remain visible and not be hidden by the main window.

Because customers may not understand window management conventions, each help window should contain an explicit button at the end of the help labelled 'Close Help', 'Close window' or simply 'Close'.

Dialogue boxes should appear as an additional window on top of the main window. They should be modal and should not be able to be hidden by the main window.

Error messages

Error messages should be absolutely clear as to what the customer should do to solve the problem. Messages such as 'some fields are filled incorrectly' are too generic, and messages with codes are too obscure. A good error message should be unambiguous about where the problem has occurred and what is the correct course of action.

Messages should be written in plain English and avoid system-oriented terms (such as 'ODBC Drivers error'). In particular, do not show generic

'404 File not found' error messages when a customer mistypes a URL at your site: configure your web server to present a customised message. This message should include a description of the problem and a description of what to do next (for example, you could provide a search function to help the customer locate the missing page).

Ideally, customers should be notified immediately if they enter an unacceptable value into a field. The error message should help the customer correct the field; for example, rather than say 'invalid telephone number' the error message should say 'The telephone number must include the STD code, e.g. (020) 7240 3388'. In practice, many web forms cannot be validated until the entire form is submitted. In these instances, re-display the form after submission with the unacceptable values highlighted (e.g. in yellow or reverse video).

Where possible, avoid errors by giving examples of correctly completed fields (for example, with a tool tip or balloon text or a text label just below the entry field). If the field refers to a currency value (e.g. 'How much would you like to transfer to your deposit account?'), include the currency symbol (e.g. £ or $) next to the field in a non-editable format so that the customer only types the amount.

Arrange objects and labels on the screen

We have now chosen the screen objects that we need and provided labels for them. Now we need to arrange the objects on the screen so that they create a usable page.

Arranging the objects on the screen properly is essential for an effective web site. Poorly organised screens can greatly hinder the customer's work, regardless of how usable the individual objects and labels are. Above all, information and information entry fields should be positioned according to customer expectations and to the requirements of the task.

As with any design process, there are trade-offs that will need to be made, for example in terms of efficient use of screen space versus the need to avoid clutter, and the developer's judgement will play an important part in selecting the optimal arrangement.

Grids, alignment and layout

Fitting the content of the form to a grid helps make the information more regular and improves the look of the information on the page. Alignment is important not only because good alignment guides the customer's attention through a screen, but also because bad alignment can be very distracting. Alignment may be achieved by creating vertical columns of screen objects and horizontally aligning the tops of these objects. Vertical orientation, which aids a top-to-bottom flow through screen objects, is generally preferable.

Alphabetic field alignment

Align alphabetic entry fields vertically in columns and left-justified within each column.

Numeric field alignment

If a group of entry fields are all numeric and the field lengths are different, display the information right-justified. If numeric fields contain decimal points, they should be aligned to the decimal point.

Field label alignment

When label lengths on a form differ significantly (e.g. 'Name', 'Credit card number'), right-justify the label next to the field. If label lengths are similar (e.g. 'Name', 'Age'), the labels can be left justified. Be consistent: do not mix field label alignments within any individual screen.

Some best practice guidance is as follows.

- Radio buttons/check boxes: left-justify button and box labels and align buttons/boxes vertically (apart from yes/no buttons)
- Text boxes: the text box field should be left-justified; numeric fields should be right-justified
- List boxes: left justify list boxes; labels positioned above the box should be left-justified
- Drop down/pop-up list boxes: the box itself should be left-justified; labels should be right-justified.

When controls are mixed and are arranged vertically, all objects should be left-justified (their labels should be right-justified).

Field lengths

Text fields should be large enough to accommodate the majority of anticipated entries without scrolling.

Visual balance

Balance is an issue similar to alignment as it can also be a factor in determining how well the design guides the customer through a screen; perceived imbalance can be distracting and annoying for the customer. The visual balance of a web page is important and can be assessed by drawing a line down the centre of the page. Balance the amount of white space, text and graphics on either side of the line.

For example, Figure 11.8 demonstrates a number of good design principles. Note that: [1] input fields and pull-down menus are cleanly aligned; [2] text

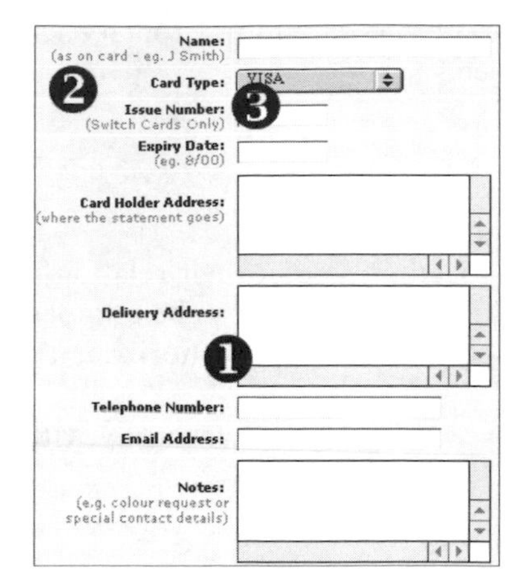

Figure 11.8 Visual balance (reproduced with permission from valuemad.com – now shopsmart.com)

below each label provides help and an example of the format required; [3] field labels are right-aligned. It would be better still if each field label was consistently aligned with the top of the corresponding input field (e.g. compare the 'Card Holder Address' field label with the label above).

Grouping of objects

In order to assist the customer's progression through the screen, objects sharing similar content should be grouped together (e.g. grouping a postcode text field with the address fields). This can be achieved in several ways:

- Spatial location: objects located close to one another will appear as a group. Adequate amount of 'white space' (i.e. space that is not occupied by screen elements) left around the grouped objects will emphasise the group boundaries and will make those elements in the group easier to read/use. There should be greater space *between* than *within* groups.
- Frames: frames are a good method of delineating groups of objects within a screen. Frames can be used to enclose a single object (in the case of radio buttons or checkboxes) or a whole section of objects. Different types of frame may be differentiated by line weight (i.e. a heavier weight for those that enclose groups of objects than for those which are used for single objects). Do not, however, enclose a single list box, text box or command button within a frame.
- Similarity: similar objects located next to one another will be perceived as a group. For example if a set of radio button objects is located together, they may be perceived as a group.

Grouping can also be used to signify dependencies between objects by positioning an object to the left or above another object. Note that objects should not touch a window or group frame, objects should not touch each other nor should labels touch frames or objects.

Best practice

- Align objects accurately, both horizontally and vertically.
- Group related components of the page together, so that their relation is clear to the customer. Objects of a similar kind will appear to be grouped if they are close to each other and when they are aligned horizontally and vertically.
- Visually separate groupings horizontally with space, and vertically with space or a horizontal rule.

Sequence/order of objects

Fields that logically go together should be adjacent. Sequence fields according to expectations: even minor decisions, such as asking for a post code before rather than after the city, can confuse customers who are accustomed to a different way of doing things.

Where there is a clear data entry/selection sequence required in order to complete the task, objects should be ordered accordingly (from left-to-right and top-to-bottom). The tabbing (between objects) should follow this sequence of objects. If there is an existing paper form that will be the source for input in the new site, try to arrange the on-line screen form in the same order.

The <TAB> key should move the customer to the next logical field: for example, in a loan application form with two-columns (one column for the applicant's details and one column for the partner's details) the <TAB> key should first move the customer through the applicant's details, then back to the partner's details (see Figure 11.9). Note that the <TAB> key moves the customer logically through the screen, not just to the next horizontally-adjacent field.

Where there is no clear sequence or paper-based equivalent, order the objects according to their associated input device; that is, try to ensure that objects which require the use of a keyboard are located together and objects which primarily require the use of a mouse are located together. The navigation through the screen should minimise the number of times the customer's hand has to travel between the keyboard and the mouse.

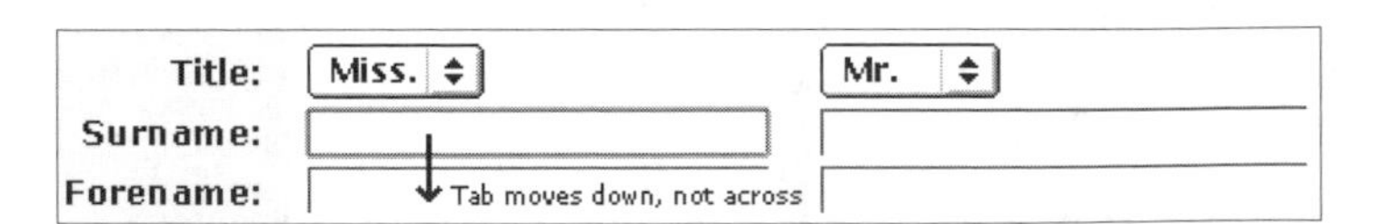

Figure 11.9 Correct tabbing order

Spacing

Ensure a minimum amount of clear space around form elements, to aid clarity and usability.

Sizing

All interface widgets – buttons, pull-down menus, check boxes, radio buttons, etc. – should have the same size throughout the form. Align items to a grid. Use the grid to adjust the sizes and positions of elements across rows. Extend short elements to begin and end on grid boundaries; allow long elements to span multiple grid units.

Buttons should be the same size as each other. Unfortunately, buttons rendered in web browsers change size depending on the length of the text in the button. Buttons should be given an approximately even size, for example by adding non-breaking spaces as part of the button name.

Arrangement

The flow of the customer's task should be reflected in the layout of the components of the screen. The actions usually performed first should be at the top and the left of the form. Keep adjacent actions close together on the screen.

Scrolling

Vertical scrolling

Although customers are familiar with the notion of vertical scrolling, the information that they see first will have a stronger impact. Consider this when designing each page in the site and make sure that the first screen of each page contains the most important information (such as navigation and location cues). Screen size will vary depending on the customer profiles but at the time of writing (2002) the most prevalent screen size is 800 pixels by 600 pixels.

Horizontal scrolling

In contrast, most customers are unfamiliar with horizontal scrolling. So you should ensure that the screens can be shown on a 800 x 600 monitor without the need for horizontal scrolling.

How to present the information

Once the screens have been laid out and tested for usability, the screen designs should be specified in a style guide. The style guide should show the design of specific screens for two or three key tasks. The developers can then use these screens as templates for other screens.

The style guide should cover the full range of user interface issues. It should:

- describe a set of general principles for good user interface design;
- provide specific guidance on interaction design, including the navigation framework for the whole site, the use of mouse buttons and keyboard shortcuts, and the design of menus and windows;
- provide general standards on the visual design of the web site, for example, fonts, colours and layout.

Where possible, include some examples of good and bad practice.

Summary

- This chapter describes how to make the leap from the information architecture to individual screen designs.
- First, transform the functional objects into screen objects. The chapter contains a flow chart to help you make the correct choice.
- Next, select labels for each screen object to ensure consistency and to maintain a logical flow between items.
- Then create any additional screen items that are needed, such as window titles, command buttons, navigational cues, prompts and additional (or pop-up) windows.
- Finally, arrange the objects and labels on the screen so that they create a usable page.
- Once the screens have been designed to support the key tasks, document the design in a style guide.

Further reading and weblinks

Books and articles

Cooper, A. (1995) *About Face: The Essentials of User Interface Design*. Foster City, CA: IDG Books.

Horton, W.K. (1994) *The Icon Book: Visual Symbols for Computer Systems and Documentation*. New York: John Wiley & Sons.

Johnson, J. (2000) *GUI Bloopers*. London: Morgan Kaufman Publishers.

Lynch, P.J. and Horton, S. (1999) *Web Style Guide: Basic Design Principles for Creating Web Sites*. New Haven, CN: Yale University Press. Also available online at http://info.med.yale.edu/caim/manual/.

Mullet, K. and Sano, D. (1995) *Designing Visual Interfaces: Communication Oriented Techniques*. Englewood Cliffs, NJ: Prentice Hall.

Travis, D.S. (1991) *Effective Color Displays: Theory and Practice*. London: Academic Press.

Williams, R. (1994) *The Non-Designer's Design Book*. Berkeley, CA: Peachpit Press.

Web pages

Instone, K. (2002) 'Breadcrumbs'. http://keith.instone.org/breadcrumbs/. A site devoted to questions and research issues surrounding breadcrumbs.

ISYS Information Architects (1999) 'Interface hall of shame'. http://www. iarchitect. com/mshame.htm. 'The Interface Hall of Shame is an irreverent collection of common interface design mistakes. Our hope is that by highlighting these problems, we can help developers avoid making similar mistakes.'

Mok, C. and Zauderer, V. (2001) 'Timeless principles of design: four steps to designing a killer Web site'. http://www.webtechniques.com/archives/1997/04/mok/. Describes ten general guidelines for developing a graphical user interface.

Nielsen, J. (2000) 'Drop-down menus: use sparingly'. http://www.uscit.com/alertbox/20001112.html. 'Drop-down menus are often more trouble than they are worth and can be confusing because Web designers use them for several different purposes. Also, scrolling menus reduce usability when they prevent users from seeing all their options in a single glance.'

Travis, D.S. (2000) 'Ten tips to make your web site usable'. http://www.system-concepts.com/articles/tenwebtips.html. This article reviews usability problems introduced by 'impressive' visual design.

12 Evaluate usability

All software undergoes *functional* testing, where the system is examined to eradicate system crashes and other bugs. But, remarkable though it sounds, a significant proportion of software reaches your computer without undergoing *usability* testing. With web sites, the problem is greatly magnified. Why should this be?

The common objection is that usability testing takes too long and costs too much money. In fact, considered over the life of the project, usability testing *saves* time and money. This is because usability problems get fixed before release (when it is cheaper and quicker to fix them) rather than waiting for real customers to experience the problems.

This fallacy stems from the pre-eminence amongst project managers of 'time to market', the date on which the site goes live. But from an organisational perspective, a more relevant statistic is 'time to profit': the date at which the site starts generating revenue. Consider a web site that needs 30,000 registered customers to make a profit. Launching a difficult-to-use version of the site in January (see the simulation in Figure 12.1) might achieve this number of customers by October (represented by the light grey bars in Figure 12.1). But an easier to use site, launched three months *later*, could achieve this number of customers a month *earlier* (represented by the dark bars in Figure 12.1). The point is that an earlier time to market does not guarantee greater profitability – and a 'time to profit' measure may be a more useful statistic.

A second objection is that many development teams simply do not know how to measure usability. Without denying its importance, many developers consider usability to be a soft measure that cannot be quantified, like 'brand value'. In fact, usability can be measured like any other engineering attribute. The key, as with most steps in this book, is to begin by taking the customer's perspective.

Why evaluate web sites from a customer's perspective?

There is always someone willing to give you comments on your web site. Often these people are fellow developers or designers in the same room as

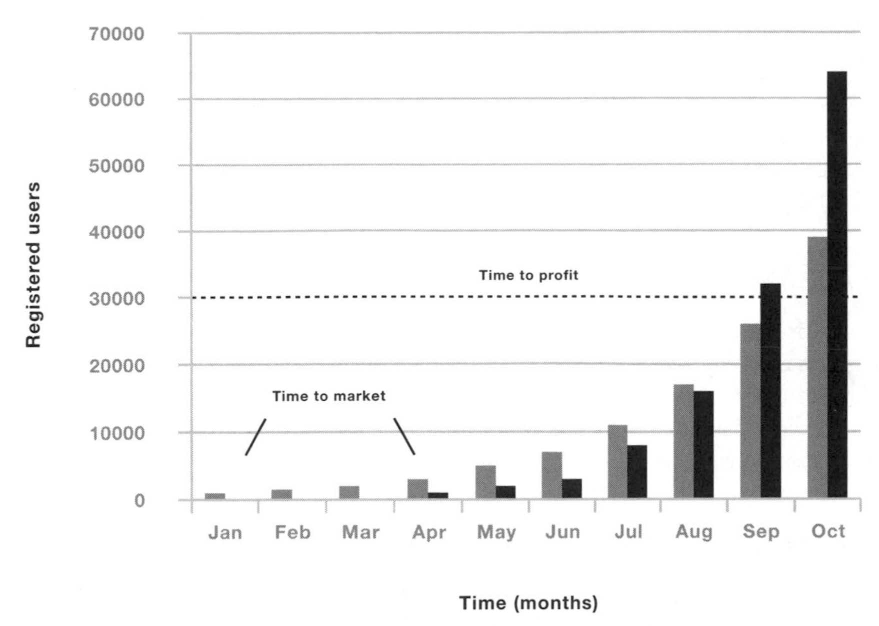

Figure 12.1 'Time to profit' under two scenarios (difficult-to-use site shown by light grey bars, easy-to-use site by dark bars)

you. These people have a lot of expertise in using and developing sites. Why not use their expertise to evaluate your site – why bother with real customers?

We talked about the problems of 'next desk design' in Chapter 4. The reasons to use real customers are to:

- find problems worth fixing;
- evaluate progress against objectives;
- choose from among design alternatives;
- determine if a web site is ready to go live;
- identify benefits to provide in future releases of the site;
- prepare to advertise benefits;
- find out how to staff and train support groups;
- find out how customers will rate your site relative to the competition.

We now consider each of these problems in turn.

Find problems worth fixing

The only problems worth fixing are those that your customers have. Peer review *is* important and should certainly be part of your quality control process. But it is insufficient for you to spot your customer's usability problems. Your fellow designers may identify usability 'problems' that your customers just do not care about. There is no point spending time fixing things that make your fellow designers unhappy but that have no impact on the way real people perceive your site.

Evaluate progress against objectives

Without customer data it becomes very difficult to measure the performance of your web site against the key performance indicators you have set. Key performance indicators, by their very nature, are based on customer data. Regular testing with customers will help you measure your progress against these objectives.

Choose from among design alternatives

How often have you sat around in a design meeting with your colleagues trying to decide between two or more different versions of a home page? How much more valuable would it be if you could have your customers choose for you? Asking customers also has the added advantage that the choice is the right one!

Determine if a web site is ready to go live

Sometimes you may have found yourself discussing changes to a design and wondering how long it will be before the site is ready to go live. There always seems to be one more thing that needs to be changed or modified before release. Testing provides you with the customer data you need to make that decision: once the key performance indicators have been met, it's ready.

Identify benefits to provide in future releases of the site

Participants are rarely happy just giving you answers to your standard questions. You will find that they have lots of comments on your site – including requests for functionality that you never intended to include. In a very real sense, usability testing brings us full circle, back once again to requirements capture. This doesn't mean that your requirements capture was incomplete; it simply means that people find it much easier to comment on what they would like to see in a product when they are able to use it. It may be too late to include this functionality in your existing web site, but you now have an idea of the kind of benefits to include in a future release.

Prepare to advertise benefits

You should already have some idea of the precise benefits that customers will be getting from your site. But there may well be issues that you have missed that will reveal themselves during testing. For example, during a usability test of some software I worked on a few years ago, experienced users were jubilant about one aspect of the software that we had not even realised was beneficial. With the new design, customers were able to move windows around on the screen. This may not sound like much of a benefit until you realise that in the previous release, the windows were in fixed positions on the screen. The customers loved it, and we had discovered a key benefit we could use to advertise in the product's marketing literature.

Find out how to staff and train support groups

During testing you may uncover some usability issues that you cannot or do not want to fix prior to release. Usability testing will not just help you identify these problems, it will also show you how to explain the solution in the customer's terms. This becomes a skeleton script that you can then pass to support groups. And once you know the type of problems that customers experience with your site, you also know the kinds of skills you need to recruit in your support staff.

Find out how customers will rate your site relative to the competition

It is just not possible to design a site that is 100 per cent free of usability problems. With a site of any complexity, there will always be *some* aspect the site that *some* customers will find problematic. So rather than spend time gold plating the site, it makes more sense to simply test it against the competition. In the land of the blind, the one-eyed man is king. Your site needs to be good enough and no better. Eventually, your site will be tested against the competition whether you like it or not. It makes sense to do the testing on your terms so that you can make changes before it goes live.

Keys to successful evaluations

In any usability evaluation programme, there are six important factors:

- Start early: use phase containment to minimise risk
- Take the customer's perspective
- Aim to find problems, not fix them
- Track defects through to completion
- Keep developers informed
- Check the test is worth it.

Start early: use phase containment to minimise risk

A usability problem introduced early in the project – for example, at the requirements phase – can cost 50–200 times as much to fix later in the project compared to the phase within which it was introduced. This is shown schematically in Figure 12.2.

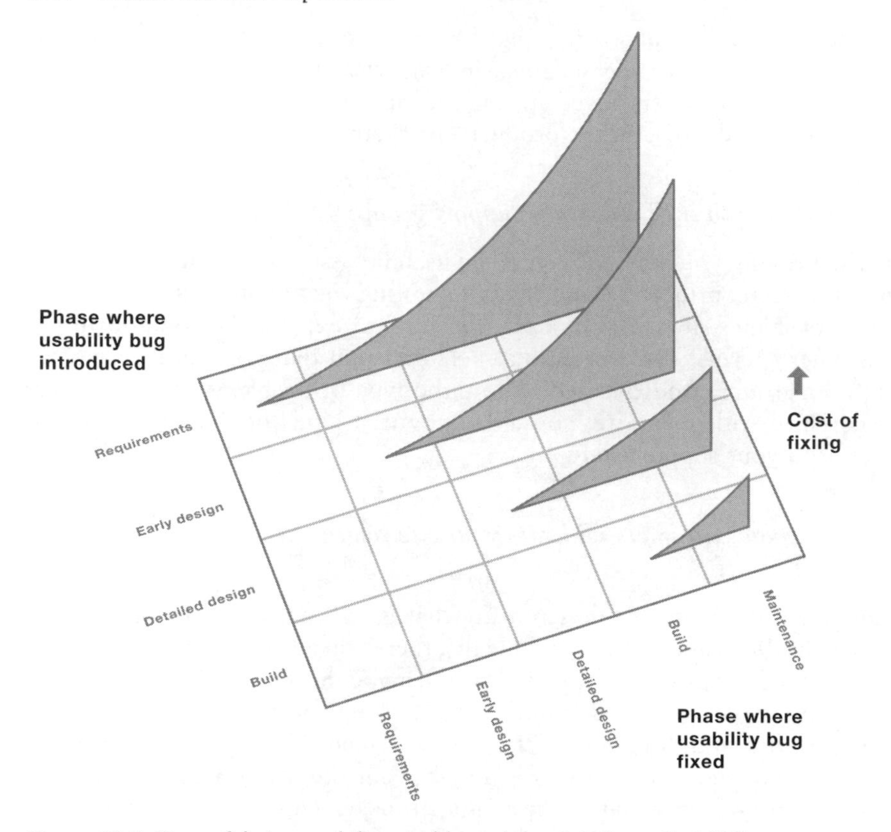

Figure 12.2 Cost of fixing usability problems (after McConnell, 1998)

This figure demonstrates that there is a real cost-benefit in detecting usability problems in the development phase where they are introduced. Allowing the usability problem to survive into the next development phase increases the costs of fixing significantly. This means there should be a usability test towards the end of each development phase so the usability problems can be found and fixed.

Take the customer's perspective

There is a real temptation to review the web site from one's own experience. But to review the site adequately, you need to take the perspective of the customer – either by involving customers in the review of the site or by trying to put yourself in the shoes of the customer. The reason for this is that the customer will have a series of preconceptions when seeing your site for the first time. Since you probably look at your site every day, it is very hard to imagine the initial responses of a new customer who has never seen your site before.

Aim to find problems, not fix them

One risk with usability reviews is that there is a tendency to try to fix the problem immediately. Without control, the usability review can degenerate into a design meeting. I have even experienced the same effect in a usability test, where a developer wanted to start work on fixing a problem as soon as the first participant has left the room. Of course, the usability problems need to be fixed but this is not the place to do it. The purpose of the evaluation is to find problems.

Track defects through to completion

Usability defects should be treated just like any other functional bug and tracked accordingly. Bugs should be entered into a bug tracking database and an individual should take ownership to close out the defect.

Keep developers informed

Developers get paid for writing code and fixing bugs. They need to schedule their time to fix the problems that are found and should therefore be informed of the problems that are identified.

Check the test is worth it

Before carrying out a test, it is worth estimating the benefits of testing. The following equation provides a simple method for carrying out this estimate.

$$\text{Cost/Benefit} = ((N \times F_{later}) + MC + LS) - (C + (N \times F_{now}))$$

where

N = Number of defects likely to be found: compute from

$$\text{UsablityProblemsFound}(i) = N(1-[1-p]^i)$$

where i is the total number of test participants, n is the total number of usability problems in the interface and p is the probability of finding any single problem with any single user.

F_{later} = Cost to fix a defect later: use the following multipliers: if the site is currently in the requirements phase, the cost to fix the defect after release will be 60–100 times the current cost; if the site is currently in the design phase, the cost to fix the defect after release will be 10–18 times the current cost.

MC = Maintenance costs because of defects: include costs for support or helpdesk personnel, costs of getting developers involved, amount of support time to handle the defect and the frequency of helpdesk calls associated with the usability problem.

LS = Lost sales because of usability defects: get estimates from marketing or make a conservative guess.

$C =$ Cost to conduct the evaluation: include the costs of usability personnel for planning the test, carrying it out, analysing the data and reporting the results. Add the direct costs of paying users, renting laboratory space and buying consumables (such as videotapes).

$F_{now} =$ Cost to fix a defect now: include the cost for developers and management to make the changes and any material costs associated with the change. Alternatively, use the following heuristic: the cost of making a change in the requirements phase is 1 unit; in the design phase, 1.5–6 units; and 60–100 units after release.

Which method?

A number of methods exist for evaluating the usability of your web site. The main distinction is between methods that involve customers in the evaluation (usability testing) and methods that solicit comments from domain or usability experts (usability inspections).

The flow chart in Figure 12.3 can be used to choose a method.

The flow chart shows that the first issue to decide is whether or not you will test with customers. To be truly customer-centred, a web site must undergo testing with customers at some stage during development. Indeed, this is a requirement of ISO 13407 (the usability standard around which this book is

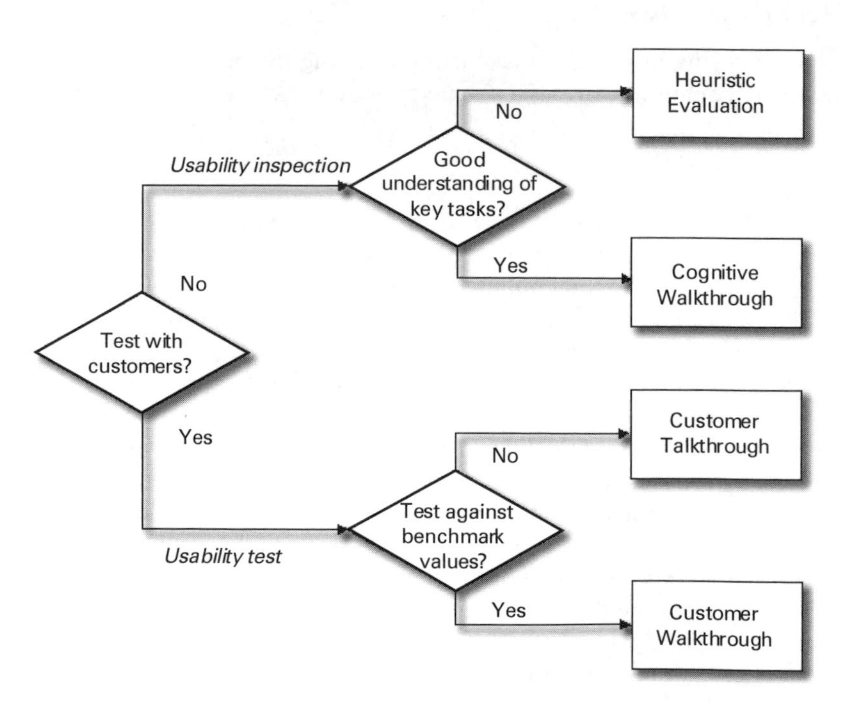

Figure 12.3 Flow chart for choosing a usability evaluation method

based). However, as part of an evaluation programme, there are sometimes good reasons to ask usability or domain experts to carry out a usability inspection.

Conditions under which you might choose an inspection over a test include when:

- You are working to parsimonious time or budget constraints.
- You just want to detect 'usability bloopers'.
- You want to identify and fix the main usability problems prior to a usability test (to prevent test participants getting thwarted by the 'obvious' problems).
- You want to make sure all the pages on your site share a consistent style.

However, usability inspections should never be the only usability evaluation method you use. This is because:

- They make critical assumptions about customers and usage that may be wrong.
- Experts vary in their quality (and in the type and number of usability problems that they find).
- It is hard to prioritise the problems that are found.

This means that every e-commerce project, no matter how small, should usability test with customers. Usability inspections should be seen as a complementary method and not as an alternative to testing.

Usability inspections

A number of methods exist for carrying out a usability inspection. In this section, we review two of the most common and most effective: the heuristic evaluation and the cognitive walkthrough.

Heuristic evaluation

With a heuristic evaluation, the usability expert reviews the user interface against a set of best practice principles. These principles are often expressed as slogans, such as 'Speak the user's language'. Many principles exist. One of the oldest (and still valuable) collections is *Guidelines For Designing User Interface Software,* by Sidney Smith and Jane Mosier published in 1986 for the US Air Force. With 944 guidelines this document remains the largest collection of publicly available user interface guidelines in existence.

In 1990, Rolf Molich and Jakob Nielsen[1] carried out a factor analysis of 249 usability problems to derive a set of 'heuristics' or rules of thumb that would account for all of the problems found. Nielsen[2] further revised these heuristics, resulting in the ten guidelines listed in Box 12.1.

The next step is to carry out the evaluation. To get adequate coverage, use around three to five different experts. This is because different experts tend

Box 12.1 Molich and Nielsen's heuristics

Visibility of system status: the system should always keep users informed about what is going on, through appropriate feedback within reasonable time.

Match between system and the real world: the system should speak the users' language, with words, phrases and concepts familiar to the user, rather than system-oriented terms. Follow real-world conventions, making information appear in a natural and logical order.

User control and freedom: users often choose system functions by mistake and will need a clearly marked 'emergency exit' to leave the unwanted state without having to go through an extended dialogue. Support undo and redo.

Consistency and standards: users should not have to wonder whether different words, situations, or actions mean the same thing. Follow platform conventions.

Error prevention: even better than good error messages is a careful design that prevents a problem from occurring in the first place.

Recognition rather than recall: make objects, actions, and options visible. The user should not have to remember information from one part of the dialogue to another. Instructions for use of the system should be visible or easily retrievable whenever appropriate.

Flexibility and efficiency of use: accelerators – unseen by the novice user – may often speed up the interaction for the expert user such that the system can cater to both inexperienced and experienced users. Allow users to tailor frequent actions.

Aesthetic and minimalist design: dialogues should not contain information that is irrelevant or rarely needed. Every extra unit of information in a dialogue competes with the relevant units of information and diminishes their relative visibility.

Help users recognise, diagnose, and recover from errors: error messages should be expressed in plain language (no codes), precisely indicate the problem, and constructively suggest a solution.

Help and documentation: even though it is better if the system can be used without documentation, it may be necessary to provide help and documentation. Any such information should be easy to search, focused on the user's task, list concrete steps to be carried out, and not be too large.

Table 12.1 Adjective/noun template for classifying usability problems

Adjective		*Noun*	
ambiguous	inconsistent	accelerator	information
cluttered	misleading	action	item
confusing	missing	animation	label
contradictory	non-intuitive	button	layout
distracting	non-standard	character	menu item
excess	obscure	colour use	message
faulty	poor	concept	metaphor
hard to find	redundant	control	object
hidden	slow	convention	option
illogical	superfluous	design element	order
imprecise	system-oriented	dialogue	phrase
inadequate	unnecessary	error message	sequence
inappropriate	vague	exit path	step
incomplete		feedback	structure
OTHER:		format	terminology
		functionality	undo/redo option
		graphic	window
		help	word
		OTHER:	

to spot different problems: Nielsen found that most evaluators get the easy problems but very few evaluators find the difficult problems. In addition, evaluators who do not otherwise find many problems find some of the hardest problems. Research has shown that using one evaluator will find about one-third of the problems; to get 75 per cent of the problems requires five evaluators.

Note also that experts should be independent of the design team and have some knowledge of the domain, as well as the heuristics that will be used.

Each expert uses the prototype and notes down the problems that he or she finds. People differ in their precise approach. Some experts like to go through the interface screen by screen noting the problems as they find them; other experts use the web site to carry out realistic tasks and then write down the problems in task sequence. One technique that may help you rapidly identify problems is to use an 'Adjective/noun' template, of the kind shown in Table 12.1.

For example, assume that while reviewing a gardening e-commerce site you discover that the icon for 'Buy an item' is a wheelbarrow rather than the more conventional shopping trolley. You could note this by selecting 'non-standard' in the adjective column and 'graphic' in the noun column. The precise terms you use are not so important (for example, you might also classify it as a 'non-intuitive button'). What matters is that you note the problem quickly and move on. Later, you will add more detail for this problem. Box 12.2 gives you some tips on estimating the number of usability defects still left in the web site.

Box 12.2 Defect statistics

How many problems did you find? Once you have collected together your list of usability problems from your three to five evaluators, you can just add them up, removing all the duplicates. A more interesting question is, 'How many problems remain?' This may seem impossible to answer, but if you used separate evaluators you can use a statistical technique known as defect pooling to estimate how far you have progressed (from S. McConnell, *Software Project Survival Guide*, Redmond: Microsoft Press, 1998).

For example, assume we used two usability experts as evaluators (E_1 and E_2). Each of these experts evaluated the interface independently of the other. To carry out a defect pooling analysis we need to note down how many usability problems each expert identified and how many duplicate usability problems were found. Assume that E_1 found 62 usability problems and E_2 found 47 usability problems. Further, assume that 36 of the usability problems were found by both experts ($E_{1\&2}$). We can then estimate the total number of unique usability problems (UP_{unique}) by the following equation:

$$UP_{unique} = UP_{E_1} + UP_{E_2} - UP_{E_{1\&2}}$$

In our example, this resolves to:

$$UP_{unique} = 62 + 47 - 36 = 73$$

The total number of defects can then be estimated using the following formula:

$$UP_{total} = \frac{(UP_{E_1} \times UP_{E_2})}{UP_{E_{1\&2}}}$$

In our example, this resolves to:

$$UP_{total} = \frac{(62 \times 47)}{36} \approx 81$$

This suggests that there are still eight usability problems (81–73) waiting to be found.

If, as we recommend, you use more than two evaluators you can still carry out a defect pooling analysis. You do this by waiting for each expert to complete the analysis and then you divide the experts into two teams. Remember to remove any duplicates from each 'team' before you do the analysis, otherwise you may underestimate the number of usability problems still to be found.

Cognitive walkthrough

The cognitive walkthrough is a semi-formal method to guide a usability inspection. Unlike a heuristic evaluation, where the evaluator can simply browse the web site and make a note of the usability problems, with a cognitive walkthrough the evaluator follows a structured process (see Appendix 9).

The structured process is based around the steps needed to complete a particular task. Indeed, the first step in the walkthrough process is to identify the atomic actions that the customer needs to carry out to complete the task. These actions represent the ideal route – the 'happy path' – that the developers expect the 'ideal' customer to take. For example, consider the following task:

> A novice user of Internet Explorer wants to change her default home page from msn.com to bbc.co.uk.

We can work out the precise actions needed to do this either by struggling with the software and working it out for ourselves or by asking the developers. A third alternative is to get the actions from the 'Help' system. For this example, there are four actions:

1 Go to the page you want to appear when you first start Internet Explorer.
2 On the 'Tools' menu, click 'Internet Options'.
3 Click the 'General' tab.
4 Under 'Home page', click 'Use Current'.

Now that we know the happy path, we can start the walkthrough proper. To do this, we ask a series of questions before, during and after each step of the process to check if the actions are realistic. These questions are:

- Will the customer realistically be trying to do this action?
- Is the control for the action visible?
- Is there a strong link between the control and the action?
- Is feedback appropriate?

The word 'control' here means the icon, button or widget that the customer needs to click on or activate. Each of these questions uncovers a different type of problem with the user interface.

Will the customer realistically be trying to do this action?

Some systems make unrealistic demands on users. For example, early ATM cash dispenser machines had a happy path something like this:

1 Enter ATM card
2 Enter PIN
3 Select 'Cash withdrawal'
4 Enter amount

5 Wait for cash
6 Wait for card

The problem was the last step in the task list. Some customers tended to walk away from the cash machine before their card was returned. The customer had completed his or her goal once the cash had been received (at step 5) and it was therefore not realistic to assume that the customer would do step 6 (that is, wait for the card to be returned).

As another example, when car alarms were first introduced, users were expected to disable the alarm before opening the car door. People forgot to disable the alarm since the realistic goal was to open the car door.

One technique for identifying unrealistic actions is to imagine how a customer might describe the steps in a task. For example, imagine purchasing an item by mail order over the telephone. If you asked a customer to specify the important steps, what response might you get?

Buying a product in this way has four important steps. First, you specify the product. Next, you check the total price. Third, you provide a delivery address. Finally, you provide payment details, such as a credit card number.

Adding any extra steps to this process, or re-ordering the steps in this process, are likely to cause difficulties. So if a web site tries to change this task by insisting that you register before you can place an order, this step will appear unnatural and will cause customers to hesitate. Similarly, if you ask customers for a credit card number before you tell them the price, some of them will bail out of the purchase.

So this question helps uncover mismatches between the designer's view of the customer's task and the customer's view of the task. Identifying these mismatches is important because when a customer stumbles across a mismatch like this at a web site, the customer perceives the site as illogical, obtuse and unprofessional.

Unrealistic actions are hard to learn, and even when learnt they are simple to forget. Therefore, if a problem is identified, it makes sense to eliminate that step (for example, manufacturers of car alarms made a single control to simultaneously disable the alarm and open the door) or put the step somewhere else, where it will not affect the flow of the task (for example, later ATM designs swapped steps 5 and 6 around). In summary, this step finds problems if:

- The customer will not expect to have to carry out this action at this point.
- It is not obvious that the action can be carried out at this point.
- The action is unnatural because customers are unlikely to have the knowledge or experience needed to do it.
- Experience with other systems may mean that the customer is expecting to do a different action at this step.

Is the control for the action visible?

The next question we ask helps us work out if customers can see the control for the action – not do they understand what it means, but simply 'Is it visible?' For example, a user that wants to arrange the icons on the Windows desktop in alphabetical order needs to select 'Arrange icons→by name' from a context-sensitive menu. This type of menu may not be appropriate in all situations. Similarly, if a menu command is buried many levels within a menu system then this question will flush it out.

This type of question also helps identify problems with icons that lack a text label and are not easily recognisable without the label. One example comes from the OXO home page (www.oxo.com). This page contains only icons. To understand what the icons mean the customer needs to mouse over the icon and read the hint text that appears.

A further example of hidden controls is hypertext links that are not underlined. The customer can find the controls – by 'minesweeping' over the display – but they are not visible on the screen. In summary, this step finds problems if:

- The control is hidden (e.g. context-sensitive, keyboard-only).
- The control is buried too deep within the menu or navigation system.
- The control for the action is non-standard or unintuitive.
- The control for the action does not have a label (and it needs one).

Is there a strong link between the control and the action?

This question helps uncover problems with the control itself. Perhaps the label for the control is ambiguous, or perhaps there are other controls that look more likely to help customers achieve their goal.

One example comes from a car park machine at Stuttgart airport. I used this machine over a period of 18 months while on assignment in Germany. The first time I arrived at the airport and wanted to leave the car park, a barrier and the control shown in Figure 12.4 confronted me. The barrier has a slot at the top and two buttons, a green one on top and a red one below. Which button would you press to lift the barrier?

Figure 12.4 Car park control, version 1.0

Convention dictates that you would press the green, upper button. In fact, this appeared to do nothing. Assuming that the barrier was broken, I pressed the red button, thinking that this would put me in intercom contact with the car park attendant who could then open it for me. To my surprise, the red button lifted the barrier.

Clearly, I was not the only person to experience this difficulty. When I returned some weeks later, the design had been upgraded. As well as a large sign (shown on the left of Figure 12.5) showing the correct button to push, some 'help text' had been added to the system (in both German and English) saying 'Please press the red button'. To emphasise this, the designers had even printed the word 'red' in red ink. Which button would you press now?

Figure 12.5 Car park control, version 1.1

As if to prove that customers do not read documentation, when I returned some weeks later, the design had been changed again (Figure 12.6). Presumably, customers had not been using the help system. The hint text, now flapping in the breeze, had been almost discarded and the design had been changed to include a series of (red) arrows, indicating which button to press.

Figure 12.6 Car park control, version 1.2

Nevertheless, the link between this control (a red button) and the action ('open barrier') was still so weak that on my next trip to Stuttgart I discovered that the control had been changed again. This time, the red button had been completely removed, and the system had been re-wired to allow customers to press the green button to lift the barrier (see Figure 12.7).

Figure 12.7 Car park control, version 2.0

As well as demonstrating the usefulness of this step in the cognitive walkthrough process, this example also demonstrates the costs of failing to include usability as part of the design process.

This step of the walkthrough is also useful for identifying a hasty time out (a step that does not allow time for the customer to select the appropriate action) or actions that are physically hard to do (such as requiring three keys to be pressed on the keyboard simultaneously). In summary, this step finds problems if:

- The label on the control is confusing (for example, an ambiguous or jargon term).
- There are other controls visible at this point that look like a better choice.
- There is a time-out at this step that does not allow the customer enough time to take the action.
- The action is physically difficult to execute.

Is feedback appropriate?

Once the customer has completed the action, he or she needs some sign that the system has accepted the action. For example, immediately the customer clicks down on the 'Buy' button, the button should change to a 'pressed in' button. The web site should then give some indication about what the next step should be: for example, a new page should load.

In summary, this step finds problems if:

- Feedback is missing: customers may try to carry out the step again or give up on the task completely.
- Feedback is hidden: the feedback is easy to miss (e.g. hidden in a status bar).
- Feedback is inadequate: the feedback is too brief, poorly worded, inappropriate or ambiguous.
- Feedback is misleading: the site's response suggests that the task has been completed when it has not.
- Customers are not prompted to take the next step, or the prompt may lead them to take the wrong step.

Appendix 9 provides a step-by-step guide to help you carry out a cognitive walkthrough.

The heuristic evaluation and the cognitive walkthrough are techniques for evaluating usability without customers. We now turn to usability testing with customers.

Usability testing

Many people working in the web industry believe they know how to carry out a usability test. The common belief is that you simply put a handful of customers down in front of your web site and you listen to what they say about it. This is a bit like saying that you create a movie by asking a few people to talk in front of a camera.

It is not simply that there is a lot more to usability testing than this simple description implies. Unstructured involvement of customers can lead you to both miss and misdiagnose problems. The method I have parodied would find only the most superficial of problems.

As an example, a number of researchers equate usability testing with traditional market research techniques, such as questionnaires, surveys and focus groups. These methods try to understand a customer's motivations for using a web site. Although these data can be invaluable for marketing and branding a web site, they provide little benefit when assessing usability. This is because people often do not understand why they are doing the things they are doing, and therefore cannot tell you. Their motivations are unconscious. And even when people are in touch with their reasons and their feelings, they may not be able to express them. They do not have the words to articulate it, or words are inappropriate.

In contrast, usability testing is a tried and tested technique for identifying usability issues with a user interface. This is because usability evaluations focus on observable behaviour – what customers do (rather than what customers *say* they do).

The key feature shared by all usability testing methods is that participants are asked to sit down in front of the web site and carry out a realistic task. This means that identifying the tasks is a critical early step in the process of usability testing. You should be able to use the task scenarios you developed in Chapter 8 with only small modifications.

For example, consider a customer struggling to use a movie listings web site to decide which movie to see this weekend. The underlying reason for the customer's difficulty may be that the customer does not understand the navigation framework – the menus and terminology. But it is rare that a customer has the experience to articulate this, and those that can articulate it may choose not to – because of a fear of looking foolish. Instead, the customer will often complain about the use of colour, or the font size, or the design of icons. This information is still relevant, but fixing them will not solve the real, underlying usability problems. The point is that focusing only on what customers say – not what they do – can badly mislead the design team when trying to find usability problems. This is where usability testing shows its strengths.

Usability testing has its roots in experimental psychology, where participants are asked to carry out one or more very prescribed tasks and a statistical analysis is performed on the results. This has not been a practical approach for many organisations and different flavours of usability testing have sprouted over the last 10–15 years. Essentially, there are now two trends in usability testing:

- customer talkthroughs;
- customer walkthroughs.

As you will see, the two methods have a lot more to unite than to divide them. With both methods, participants are usually tested one at a time and are asked to carry out specific tasks with the web site. The test administrator records participants' behaviour as they use the site, either by making a video recording or by taking detailed notes. The main difference between the methods is in the number of participants that are used and in the type of data collected. Talkthroughs can use as few as five participants and aim to answer the question: 'What is wrong with this web site?' In contrast, walkthroughs tend to use around 20 participants and aim to answer the question: 'How good or bad is this web site?'

Customer talkthrough

With this approach to usability testing, each participant 'thinks out loud' as he or she uses the web site. The test administrator is usually present in the same room as the participant and encourages the participant to keep up a good flow of comments.

An example

Imagine a task at an e-commerce site where a participant has been instructed to buy a Nintendo Game Boy for a nephew's birthday present. Typically, participants are reluctant to talk during the test because they feel self-conscious, so the test administrator will ask a series of open questions. If we sat in during part of the test, it might sound a bit like this:

Administrator: What are you trying to do at the moment?
Participant: I'm trying to buy a Game Boy.
(The participant scrolls to the bottom of the page.)
A: Tell me what you are thinking.
P: I'm looking to see if I can find the word 'Game Boy' on this page – but I can't.
A: What would you do next?
P: Well I could search for it, but I don't like to do that.

A: You don't like to search?

P: I can use search but I'm not sure if the word 'Game Boy' has a space in it or not. Also, it might tell me about all the 'games' it has in stock for 'boys'. Then I'd get lost.
(The participant drums his fingers.)

A: What would you do at other sites?

P: I'd use the navigation bar.
(He scrolls back to the top of the screen and spots the navigation bar.)
Ahh, I could use the tabs at the top of the screen.
(Pauses.)
But I don't know which one to pick.

A: Keep talking.

P: Well there are two items here that both seem good choices: 'Electronics' and 'PC & Video Games'. Oh, and then there's 'Toys & Kids'. I didn't see that at first. So now there are three items and I don't know which of those to choose.

A: Go on.

P: (Pause.) I would probably pick 'Electronics'.

A: What helped you make that decision?

P: Now I think about it, 'PC & Video Games' sounds like software – the games themselves. 'Toys & Kids' still seems a good choice, but I think 'Electronics' is a slightly better description of what I want. So I'll choose 'Electronics'.

A: So what I'm hearing you say is that you think the Game Boy will be listed under either 'Electronics' or 'Toys & Kids', but 'Electronics' seems more appropriate.

P: Yes, and if I'm wrong I'll just click the 'Back' button and choose again.
(The participant clicks on the 'Electronics' tab and the next screen loads.)

This excerpt reveals a number of advantages of the talkthrough method. First, we uncover what the participant is trying to do: in this example, he is looking for a link to the item from this page and when he can't find it he expects to use a navigation bar. Second, we understand why the customer is confused: the site's navigation bar is at the top of the page and the participant has scrolled it out of view. Third, we discover something about this participant's preferred style (he seems to prefer to browse rather than search). And fourth, we discover some problems with the terminology used on the tabs.

The quality of the results depends crucially on the quality of the questions that are asked. Questions should generally be open, as in the transcript above. You should not lead participants, and the tone of the questions should be kept non-judgemental. For example, poor questions would be: 'Why did you have problems on that page?'; 'Why don't you try scrolling?'; 'Where do you think the "Electronics" link will take you?' These types of question point the participant to the right choice and can make them feel foolish.

Closed questions should be restricted to those situations where clarification is required (for example, if the participant says 'That was a bit complicated', ask 'What was a bit complicated?'). But beware the use of closed questions to get the participants to do your design for you. For example, you could say to the participant, 'So you don't like the term "Electronics". Can you think of a better term?' Of course, the participant will suggest something but the chances of this being any better than your original term are slight. Similarly, asking people which of two icons they prefer is fraught with difficulty. People will always choose one, and will even be able to rationalise their decision, but this is a poor way to do design. It is more productive to usability test the alternatives and see what people do (rather than listening to what they *say* they would do).

How many participants?

In 1993, Nielsen and Landauer[3] carried out a study to investigate the number of test participants needed in a usability test. In this study, the authors showed that the following formula gives a good approximation of the finding of usability problems:

$$\text{UsabilityProblemsFound}(i) = N(1-[1-p]^i)$$

Where i is the number of test participants, N is the total number of usability problems in the interface, and p is the probability of finding any single problem with any single test participant (in Nielsen and Landauer's study they referred to this variable as λ).

By setting N = 100 per cent and choosing an appropriate value for p Nielsen and Landauer were able to work out how many users are needed for a usability test. For example, Figure 12.8 shows the results with p = 31 per cent (that is, there is a 31 per cent chance that any single usability problem will be found with any single test participant).

The figure shows that as more participants are tested, there is a decreasing return on investment. Testing five participants identifies over 80 per cent of the usability problems, but doubling the number of test participants to ten finds only a further 13 per cent of problems. If the aim of your usability test is to find usability problems, it therefore seems pragmatic to run a few small tests with about five participants rather than run one large test with ten participants or more. This approach also fits well with the concept of phase containment where a number of tests are carried out during the development lifecycle, rather than carrying out one large test at the end of development.

But a few points need to be made about this analysis. First, it assumes that there is a 31 per cent chance that any single usability problem will be found with any single test participant. This number will only apply to a site with a fairly restricted customer base, where one customer is very much like the next. For sites with a more diverse customer base this number will be optimistic.

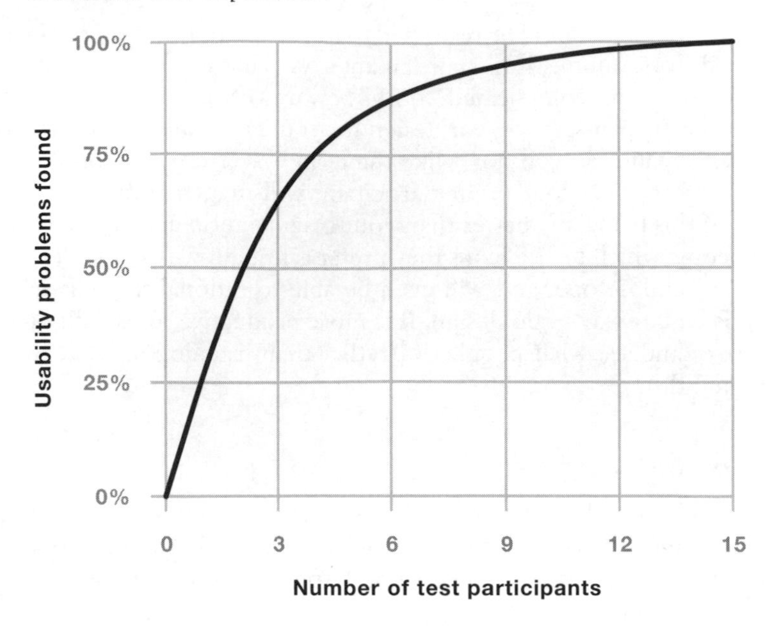

Figure 12.8 Usability test results for $p = 31\%$

To give a specific example, if the web site uses a small font size and the site is tested with young customers, the chances of detecting this as a problem may be as low as ten per cent, since most young people have normal, or corrected to normal, vision. This is fine if the site is intended only for young people, but if the site is also intended for use by older people, then the chances of any customer experiencing the problem will be much higher. In these situations, you should estimate the number of participants by reducing the value of p. Using the equation above, if we reduce p to 10 per cent, around 15 users will be needed to find 80 per cent of the usability problems. As with all usability testing, the moral is to test with representative customers.

A second issue with this analysis is that for your site, fixing just 80 per cent of the problems may not be good enough. This would be another situation that means you need to test with more participants.

Customer talkthrough: summary

The aim of the talkthrough technique is to find usability problems with the web site. It is not an appropriate method for estimating how long people take or how successful they are likely to be. This is because asking the participant to think out loud has an effect on the speed at which the participant can complete the task. And participants are more likely to complete the task successfully with this technique because the very act of questioning can lead the participant to try different things.

Customer walkthrough

You use the customer walkthrough technique when you want to know how well (or how badly) your web site is performing against benchmark values. For example, you may want to measure how long a customer takes to buy a product at the site. Or you may want to know your site's overall usability score: what are the chances that any customer will successfully complete their tasks? Or you may want to know how you perform against the competition. A customer walkthrough is the only way of checking that you have met the key performance indicators you set for the site.

Customer walkthroughs will also identify where in the process usability problems occur. For example, when you come to analyse your data you may find that participants take 15 minutes to purchase a product and that 70 per cent of their time is spent searching for the product. This makes it clear that the search process is the area that needs some attention.

A customer walkthrough is often characterised as an 'experiment'. Indeed, some of the procedures from experimental psychology are used, such as statistical analysis. But it is important to note the differences too. It is very rare that a commercial organisation has the time or resources to invest in the kind of tightly controlled, rigorous usability test that would pass muster in a university psychology department. A customer walkthrough is experimental psychology on a budget.

The main problem with the customer walkthrough technique is that the results from different participants can be quite variable. For example, one participant may complete the task in five minutes and the next participant may take 15 minutes. Some of this variability is inherent in the customer population and it is therefore appropriate that you measure it (see also the discussion of standard error in Chapter 9). But some of the variability may be introduced by inconsistencies in the way the test is carried out. To reduce variability in your own test sessions:

- Help participants relax: make them a drink when they first arrive, and describe what the evaluation entails. Emphasise that it is the web site, not them, that is being tested. Use a common routine for briefing and de-briefing participants to make sure that all test participants have the same level of understanding.
- Use at least ten participants: the precise number of participants that you need depends on the quality of the data you want to collect but ten is an absolute minimum. Averaging data with fewer participants is not very meaningful.
- Do not help participants unless they get really stuck: once you help participants, their results for that task are contaminated.
- Randomise test sessions for different customer groups: if you test two groups of participants (for example, web novices and web experts) be sure to mix them up in your testing schedule. For example, do not test one group in the morning and the other group in the afternoon or evening, because there might be other differences between those two sessions (in

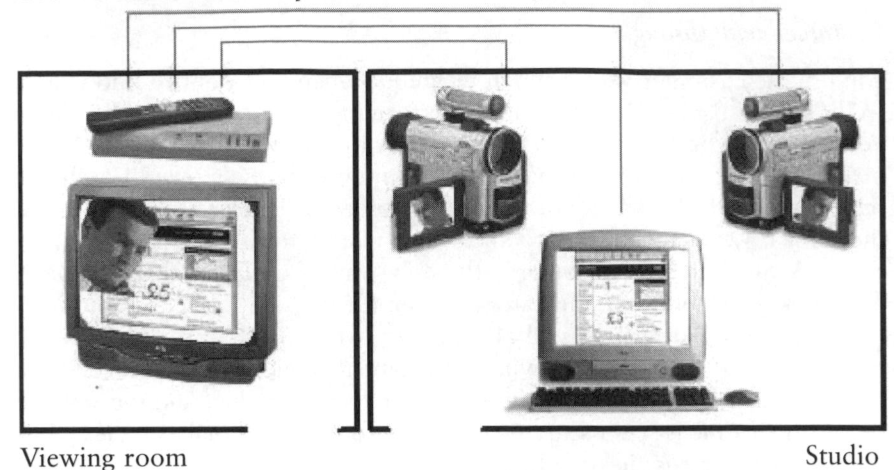

Figure 12.9 Traditional usability lab

the morning, the web server might be quicker or the room might be cooler, etc.)

Usability testing facilities

Early usability testing facilities were a hybrid between the laboratories of experimental psychologists and the viewing facilities of focus groups. So for example (Figure 12.9) the test participant would sit in the studio and be recorded by video cameras as he or she worked on the web site. The test administrator and any interested observers would sit in the viewing room where they could watch the customer either on video or through a one-way mirror.

This type of facility is still a valuable one, especially for customer walkthroughs where the test administrator leaves the participant to work alone. The video signals from the cameras and from the computer (the latter sent through a scan converter to preserve quality) are fed into a mixer that allows the administrator to derive various combinations of camera and computer views.

But not all types of usability testing need such an elaborate facility. The customer talkthrough technique requires no more than some quiet office space. However, even for this type of testing, video recordings can be useful because the test administrator can often get so engrossed in the usability test that he or she misses an important component of behaviour. One simple technique, first suggested by Gary Perlman[4], is to use a single video camera pointed directly at the screen of the computer with a mirror carefully angled to capture the facial expression of the test participant (see Figure 12.10).

With careful framing of the video image, this technique can achieve a kind of 'picture in picture' effect. The video recording of the screen image

Figure 12.10 Recording a test session with just one camera (after Perlman[4])

suffers from scanning interference (this can be ameliorated by using an active matrix liquid crystal display on the computer rather than the standard cathode ray display) but it is simple to set up and needs just one video camera.

In addition to video equipment, you will need data collection tools. These can vary in sophistication from data logging software (to measure the time of tasks and sub-tasks) to eye movement recordings (to find out where, precisely, the test participant is looking). However, for all but the most demanding usability tests, you should find that you can get by with some carefully prepared response sheets, a pencil, and a stopwatch.

Usability testing step by step

Both kinds of usability test – customer talkthroughs and customer walkthroughs – require a certain amount of planning. In this section, each of the steps is described.

Write a test plan

The first step is to write a test plan. The test plan serves as a vehicle for defining the evaluation, scheduling resources and obtaining buy-in. It also describes the customer population and the task scenarios for the test.

The test plan is typically a short document, describing the purpose of the test, the participants, the evaluation procedure and the data that will be collected. The content of these sections is described in more detail below.

PURPOSE OF THE USABILITY TEST

It is very easy to waste money on a usability test by just collecting data with no aim in mind. Typical aims might be:

- to find the good and bad aspects of the interface so that you can see how the design can be improved;
- to decide if the web site is acceptable;
- to choose between two or more alternative sites;
- to carry out an analysis of a competing web site.

PARTICIPANTS AND RESPONSIBILITIES

In this part of the test plan, you describe the profiles of the customers that you will recruit for the test. You specify how you will encourage the participants to attend (for example, with a gift or with a cash payment). You include a participant screener that defines how you will make sure you recruit the right participants. You also name the people who will carry out the test and what their responsibilities are.

EVALUATION PROCEDURE

This part of the test plan outlines the procedure itself: for example whether you will use the talkthrough or walkthrough technique. It also describes when and where the test will be carried out. You should also describe the test tasks in this section and include any questionnaires that will be used to collect opinion data once each task (or the test as a whole) is completed.

DATA COLLECTION

This part summarises the way data will be collected. For example, for qualitative data you might note down any difficulties faced by the participant, and any verbatim comments. For quantitative data, you might rate participants' success rate for each task by assigning 100 per cent to a successful completion, 50 per cent to a partial completion, and 0 per cent to a failure (see Table 9.1 in Chapter 9 for an example); you might measure the amount of time that participants take to complete each of the tasks and sub-tasks (see Table 9.2 in Chapter 9 for an example); and you might measure participant satisfaction by asking each participant to fill out a questionnaire on completion of each task.

Prepare for the test

The next step in the usability test is to prepare for the test itself. This means setting up the test facility, preparing non-disclosure agreements if required, printing participant consent forms to allow you to video record the test session, and creating any data collection sheets that you need.

Conduct a pilot study

The purpose of the pilot evaluation is to eliminate flaws in the evaluation procedure and test materials. This helps you remove any 'bugs' so that you can collect valid data during the formal usability test. This is your chance to answer questions such as:

- Does the prototype work for all the tasks?
- Is the recording equipment working properly – and is there a suitable supply of tapes?
- How long does the test take to administer?
- Where are the main usability problems likely to occur?
- Are there any ambiguities in the participant instructions or the task scenarios?

Conduct the test

The next step is to conduct the test itself, using either the talkthrough technique or the walkthrough method.

Analyse the results

We then move on to an analysis of the results. The type of analysis that you carry out will be closely linked to the key performance indicators that you set for the web site. For the talkthrough method, the data analysis will focus more on a description of each of the usability problems and suggested solutions for fixing them. You could also carry out some rudimentary analysis on the type of problems that you identified. For example, you might estimate the severity of the usability problems as 'Low', 'Medium', 'Serious' or 'Critical'. You could then produce a table identifying how many usability problems you found in each category.

For the walkthrough method, you will analyse the data in terms of effectiveness, efficiency and satisfaction. As described in Chapter 9, useful measures include the accuracy of completed tasks; the completion rate (the percentage of participants successfully completing the tasks); and the tasks completed per time unit (completion rate/mean time on task). You should also include some qualitative analysis, for example a rating scale for satisfaction – as well as any colourful comments from participants!

Tracking usability problems

Most proper software development projects contain procedures for tracking and fixing functional bugs in the software. In contrast, fewer projects track usability defects in the same way. Since it is just as important to fix usability problems as functional problems, these defects should be entered into your defect tracking system.

It makes more sense to enter the defects into your existing defect tracking system than to create a new system because then usability defects are treated in the same way as functional defects. It also makes it more likely that they will be fixed. But if your project lacks a defect tracking system, the template in Box 12.3 can be used as a start.

Prioritising usability problems

The output from your usability inspection or your usability test will be a list of usability problems with the web site. It is unlikely that the development team will have time to fix all of them. The purpose of this section is to provide you with a consistent method to assign severity definitions to usability problems.

In practice, you will probably make the estimates from your personal experience evaluating other sites. Unfortunately, there is a great deal of variability amongst experts in the assignment of severity definitions. One technique to control this variability is to ask a number of evaluators to assign a severity

Box 12.3 Defect tracking template

Index: A unique number to help you keep track of this problem.

Location: A description of where in the user interface this defect occurs so you can find the problem again.

Usability issue: A description of the usability issue. For example 'No feedback given when saving the file'.

Potential problem: The potential problem that this may cause. For example, 'User may save the file multiple times, increasing traffic over the network and causing delays for all users'.

Severity: Define this as Low, Medium, Serious or Critical.

Possible fix: Use this section to describe a possible solution to the usability problem after you have discussed the issue with the development team. For example, 'Use a progress bar to show that the file is being saved to disk'.

Cost of fixing/cost of keeping: Get an estimate from the development team of how much the possible fix would cost to implement. Also estimate how much the defect will cost the project if it remains in the system.

Owner: A named individual who is responsible for tracking this defect to completion.

Status: A description of the current status of this defect.

definition and then simply average the results. Another technique is to use the 'impact forecasting tool'.

Impact forecasting tool

Some usability problems will stop the customer in his or her tracks – the customer may not be able to find the product that he or she wants to buy. Other problems may cause only slight inconvenience, such as a typographical error. Some problems will affect only a small group of customers (such as novices) while other problems may affect all customers. Some problems will appear every time the site is used (for example, a confusing sign-in process) whereas others may affect customers only once (for example, a poorly-chosen image that customers need to learn is a button).

The three dimensions of a usability problem are therefore:

- Consequence: what effect does it have on the customer?
- Frequency: how often will customers be affected?
- Magnitude: how many customers are affected?

To estimate consequence, consider what the effect of the problem would be if it happened just once. Classify this as low, medium or high. For example, imagine that a site requires installation of a program (such as a plug-in) before it can be used properly. If the installation process failed, the consequence would be high since the site could not be used at all. In contrast, imagine a typographic error in the plug-in download instructions. If the customer can still judge the meaning of the instructions from the context, then the consequence will be low. Other examples are given below:

A *low consequence problem* is one that would cause a typical customer to:

- re-read a sentence or word;
- draw attention to poor aesthetics;
- 'Undo' an error (for example, use the 'Back' button);
- slightly delay task completion;
- feel a bit dissatisfied.

A *medium consequence problem* is one that would cause a typical customer to:

- ignore instructions, prompts or links because they are unclear or because they cannot be read (for example, the instructions disappear too quickly or they are printed in a small font);
- enter false or erroneous information (by accident);
- feel frustrated or angry;
- seriously delay task completion;
- have to look for (and find) a workaround;
- miss important functionality because it cannot be found;

- deviate from the 'ideal path' to complete the task (but still complete the task);
- access on-line help.

A *high consequence problem* is one that would cause a typical customer to:

- damage his or her computer or cause the computer to crash;
- make an error that cannot be corrected easily;
- fail to complete the task;
- leave the web site with his or her goals incomplete;
- phone the helpdesk or technical support.

To estimate frequency, judge the probability that any particular customer will encounter the problem (more than 80 per cent of the customer base, 20–80 per cent, or less than 20 per cent).

To estimate magnitude, judge how often the problem is likely to occur for those customers who experience it (daily or more frequently, weekly, or monthly or less frequently).

The final step is to use the flow charts in Figures 12.11, 12.12 and 12.13 to generate a severity definition.

For example, assume that you have identified a usability problem: the screen where the customer enters their delivery address is poorly laid out (the text input fields are not correctly aligned). This will be a 'low' consequence problem linked to poor aesthetics. We now turn to Figure 12.11.

The next question we ask is, 'How often will the affected customers experience this problem?' This problem will happen every time the customer

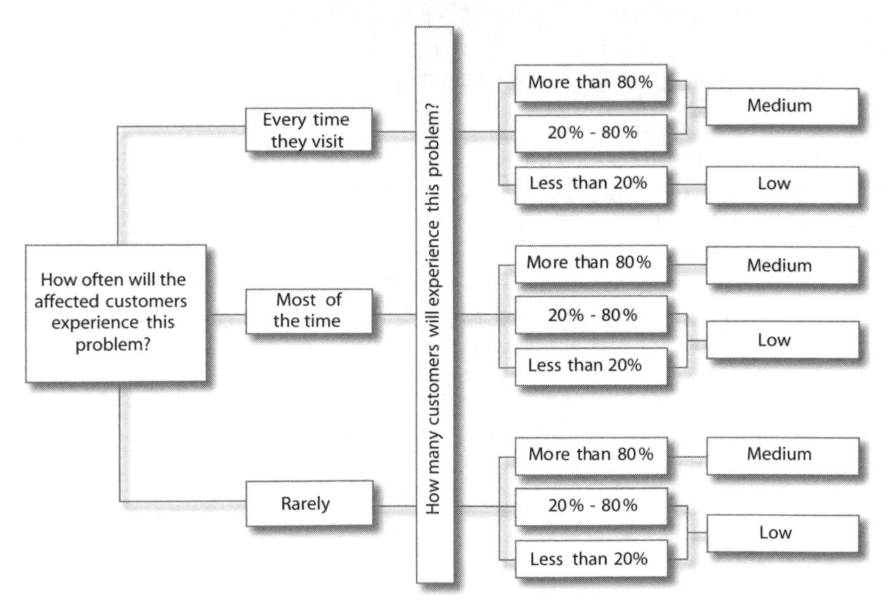

Figure 12.11 Flow chart for usability problems of 'Low' consequence

makes a purchase, but not every time they visit the site: we therefore choose 'Most of the time'. Then we ask, 'How many customers will experience the problem?' All customers will experience this problem (whether or not they can articulate it) so we choose 'More than 80%'. This provides us with a severity definition of 'Medium'.

As another example, assume that during a heuristic evaluation of a web site we find that there are two buttons at the end of a page that contains a form: 'Submit' (which sends the information to the web server) and 'Clear' (which removes all of the information that has been entered so that the customer can start again). The 'Submit' button has been designed as a complicated graphic and does not look like a button, whereas the 'Clear' button does look like a button.

We could imagine that some customers, on completing the form, will press the 'Clear' button since it looks more like a 'standard' web button than the more fancy 'submit' button. This will lead to customer frustration, since all of the address and payment details will have to be re-entered by the customer. So we classify this as a 'medium' consequence problem and then use Figure 12.12 to estimate the severity of the problem.

Once a customer has made this mistake, it is unlikely that the problem will recur for that customer so we can safely bet that this problem will occur 'Rarely'. Following the flow chart, we then ask 'How many customers will experience the problem?' Our experience suggests that most customers will experience this problem, so we choose 'More than 80%'. This gives a severity definition of 'Serious'.

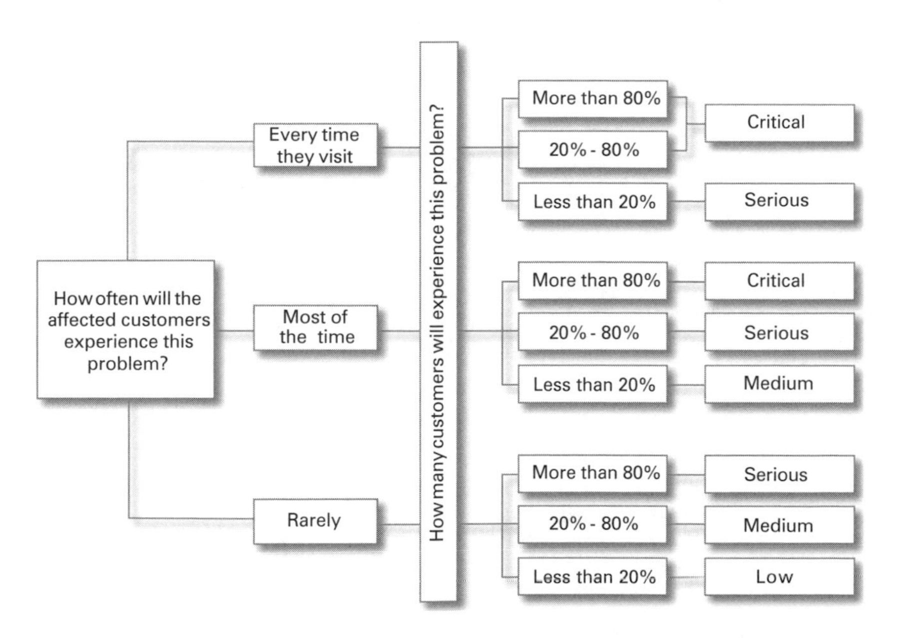

Figure 12.12 Flow chart for usability problems of 'Medium' consequence

As another example, assume that during a usability test of a mortgage application web site we discover that nearly all of our customers misinterpreted 'household income' (that is, the income of all the wage earners in their house) as their own salary. Assume further that it is difficult to change this information once the application has been submitted for processing (without going through the whole process again). So we classify this as a 'high' consequence problem and then use Figure 12.13 to estimate the severity of the problem.

You will see from Figure 12.13 that once a problem has been defined as 'high' consequence, the severity definition is inexorably 'Critical'.

Once you have the severity definitions, you need to make a decision on which ones to fix.

- Fix critical problems prior to release, even if this means delaying release of the site.
- Fix serious problems prior to release so long as doing so does not have a major impact on the schedule. Serious problems not fixed prior to release should be fixed very soon after and certainly in the next release of the site.
- Fix medium severity problems in time for the next release.
- Low severity problems can be postponed indefinitely. Fix them if there is time in the schedule and doing so will not delay other work.

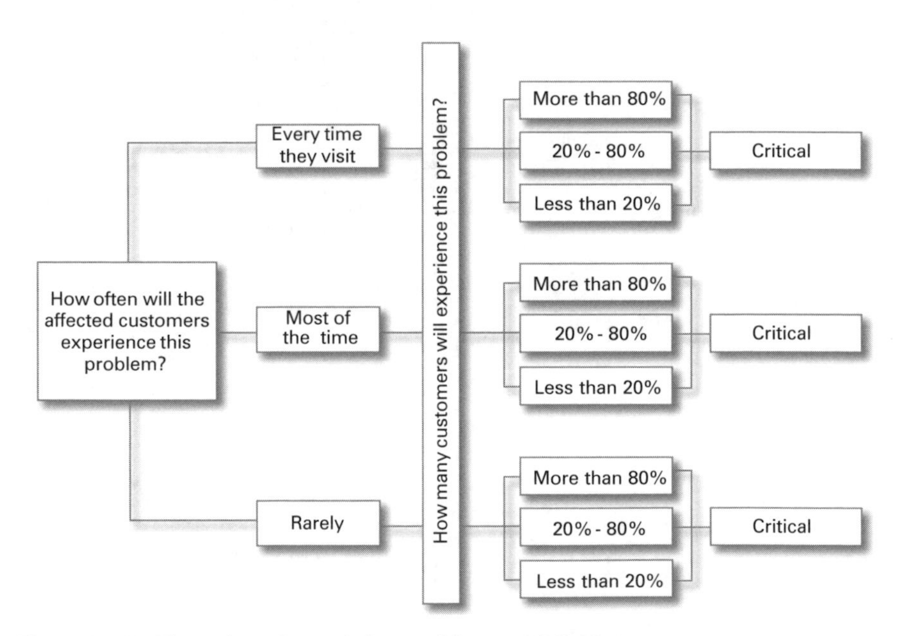

Figure 12.13 Flow chart for usability problems of 'High' consequence

How to present the information

Each person has a unique style for presenting the results of a usability evaluation. This might take the form of a written report, a presentation, or an article on your Intranet. If you are presenting the results of a usability test (rather than an inspection) you have some more options: a 'highlights' video to make the customer real, a road show or a photo album.

The type of report you produce depends upon who needs the results. Optimise the format of the report for the people who will be fixing problems. Use a 'Hot spots' report (a presentation-style report) to summarise the results to managers and all team members. Individual developers will appreciate a 'Defect Report', where all of the usability problems are listed and categorised by their severity. 'Formal Reports' are important to describe in detail what you did, so other people can use the same method in the future and benchmark their results against your own. For usability tests, a useful standard now exists – ANSI/NCITS 354 'Common Industry Format for Usability Test Reports' – that provides detailed guidance on what such a report should contain.

The main point to bear in mind for each of these reports is the potential audience. Developers, designers and managers will read your report and they have different needs.

Developers/designers

This group will probably include the very people who designed the site that you are critiquing. If you want these people to listen to what you have to say, be careful not to play the role of design police. Point out the good parts of the design. Make the evaluation tangible by describing the tasks you used to evaluate the system, and by including some good and bad examples of screens. To report on a usability test, describe the participants and frame-grab pictures from the video. Identify the Top 10 important results, with quotations from participants if you have them. Try to involve developers during the report writing process: for example ask them to help prioritise problems so they take ownership of them.

Managers

For this audience, write reports with high-level summaries. For a usability test, include lively quotes from participants. For all types of evaluation, emphasise the cost benefits (such as reduced re-work and fewer errors). Point out those results that will make the company money.

Summary

- Usability testing will save your project time and money and will ensure that your web site has a more immediate 'time to profit'.

- Web sites need to be evaluated from the customer's perspective.
- Testing should be carried out at the end of each development phase. This ensures that usability defects are trapped in the development phase in which they are introduced.
- There are two main techniques for evaluating usability: usability inspections (carried out by usability and/or domain experts) and usability tests (where representative customers use the site).
- Usability inspections are only as good as the experts used to carry them out.
- Usability tests ('talkthroughs') that aim only to find usability problems can use as few as five participants. Tests that need estimates of benchmark values ('walkthroughs') require at least twice as many participants.
- Usability problems should be formally logged and prioritised to make sure that the right problems are fixed prior to release.
- Be creative when you feed the results back to the design team and be careful to emphasise the good parts of the design, as well as the bad.

References

1 Molich, R. and Nielsen, J. (1990) 'Improving a human-computer dialogue'. *Communications of the ACM* 33, 3 (March), 338–48.

2 Nielsen, J. (1994) 'Enhancing the explanatory power of usability heuristics'. Proc. ACM CHI'94 Conf. (Boston, MA, April 24–28), 152–8.

3 Nielsen, J. and Landauer, T.K. (1993) 'A mathematical model of the finding of usability problems'. Proc. ACM INTERCHI'93 Conf. (Amsterdam, The Netherlands, 24–29 April), 206–13.

4 Perlman, G. 'Using a Mirror for Usability Testing'. http://www.acm.org/~perlman/mirror.html.

Further reading and weblinks

Books and articles

ANSI/NCITS 354. *Common Industry Format for Usability Test Reports.* Document Number: ANSI/NCITS 354-2001. Washington, DC: InterNational Committee for Information Technology Standards. (Available on-line at http://www.techstreet.com/cgi-bin/detail?product_id=918375.)

Dumas, S.D. and Redish, J.C. (1999) *A Practical Guide to Usability Testing*. Exeter: Intellect.

Nielsen, J. (1993) *Usability Engineering*. Boston, MA: Academic Press.

Nielsen, J. and Mack, R.L. (eds) (1994) *Usability Inspection Methods*. New York: John Wiley & Sons.

Polson, P.G., Lewis, C., Rieman, J. and Wharton, C. (1992) 'Cognitive walkthroughs: a method for theory-based evaluation of user interfaces'. *International Journal of Man–Machine Studies*, 36, 741–73.

Rubin, J. (1994) *Handbook of Usability Testing: How to Plan, Design, and Conduct Effective Tests*. New York: John Wiley & Sons.

Shneiderman, B. (1997) *Designing the User Interface: Strategies for Effective Human-Computer Interaction*, third edition. Reading, MA: Addison-Wesley.

Wickens, C.D. (1984) *Engineering Psychology and Human Performance*. Columbus, OH: Charles E. Merrill Publishing.

Web pages

Gordon, S. (2000) 'How to plan, execute, and report on a usability evaluation'. http://builder.cnet.com/webbuilding/pages/Graphics/Evaluation/. This article walks you through the process of setting up and implementing a usability evaluation.

Nielsen, J. (2000) 'Heuristic evaluation'. http://www.useit.com/papers/heuristic/. Jakob Nielsen's on-line writings on heuristic evaluation.

Nielsen, J. (2000) 'Why you only need to test with 5 users'. http://www.useit.com/alertbox/20000319.html. 'Some people think that usability is very costly and complex and that user tests should be reserved for the rare web design project with a huge budget and a lavish time schedule. Not true. Elaborate usability tests are a waste of resources. The best results come from testing no more than five users and running as many small tests as you can afford.'

Part IV
Step 4
Track real-world usage and continuously improve the site

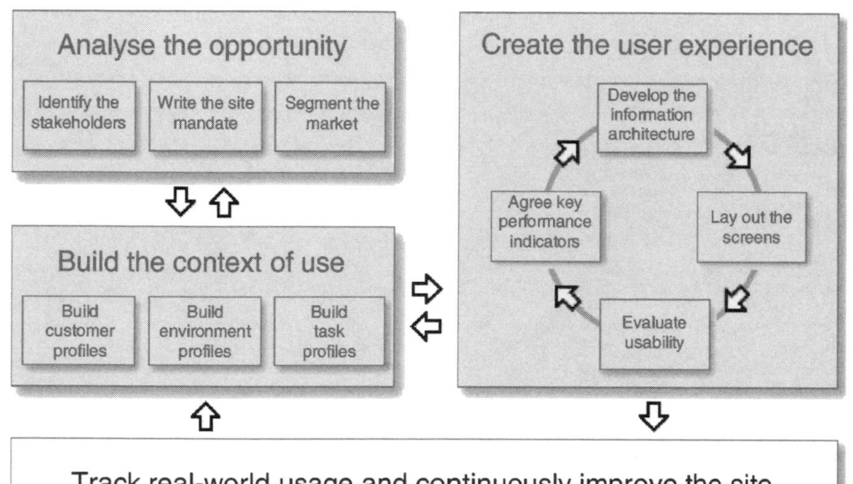

13 Track real-world usage and continuously improve the site

Introduction

It is tempting to think that, other than a few housekeeping tasks, we are now at the end of the web development process. Of course, you will need to check on the site and fix broken links, respond to the occasional e-mail and update some of the content. But effectively the design is completed – isn't it?

Unfortunately, no. To paraphrase Winston Churchill: this is not the end. It is not even the beginning of the end. This is the end of the beginning.

To maintain a winning site, you will need to change and update it as your customers change. Your customers will update their skills, their tasks and their environment. You will also need to track several key measures to keep your finger on the pulse of your site. In the same way that a change in blood pressure can indicate an impending health problem, changes in some of these statistics can point to the fact that your site needs some attention.

E-commerce health indicators

This section describes some basic statistics that you should collect and analyse on a regular basis:

- conversion rate;
- fulfilment;
- customer retention.

Conversion rate

Conversion rate is the key metric for measuring the success of your web site. Conversion rate is simply the percentage of customers who do what you want them to do as a result of using your site: this might be purchasing a product, but it could also be joining your mailing list or visiting one of your 'bricks and mortar' stores with a voucher obtained on-line.

Companies are, in general, fairly secretive about their own conversion rates but the generally accepted statistic is that the 'average' rate is between two and five per cent. Of course, this hides a lot of variation. Perhaps the best principle to take from these numbers is that conversion rates are low, and yours will probably be a lot lower than you thought.

In practice, rather than fret over whether your conversion rate is higher or lower than a competitor's, it may be more valuable to look at the 'growing room' you have in your own rate. For example, imagine that you run an e-business selling carpet slippers and your current conversion rate is 2 per cent. Your access logs show that you get around 100,000 unique visitors a week. Assume further that on average, people buy one pair of slippers at an average cost of £15. What kind of growing room do we have?

First, we can use a very simple equation to calculate our total weekly sales:

Weekly sales = Conversion rate × Average sale × Number of visitors

In our example:

Weekly sales = 2% × £15 × 100,000 = £30,000

Our conversion rate of 2 per cent shows that we have 2,000 customers per week (2,000 'bookers' versus 98,000 'lookers'). Our business analysis may indicate that this level of sales is too low. Perhaps we need to reach sales of £45,000 per week before we can start making a profit. Assuming that our carpet slippers are priced correctly, we have just two options:

- increase the conversion rate;
- increase the number of visitors.

Increasing the number of visitors can be achieved by marketing: advertisements in newspapers, magazines and television or on other web sites; and word-of-mouth recommendations or other viral marketing techniques. However, this approach can be expensive – an extensive advertising campaign may double the number of visitors but if the cost of the campaign is £500,000 then this means it costs £250 to acquire a new customer (since 100,000 new visitors will result in just 2,000 new customers).

An alternative approach is to focus on the 'lost' 98,000 visitors and try to work out how we can convert more of them into paying customers. Perhaps some customers do not want to register before shopping; or perhaps there are some who find the 'terms and conditions' so off-putting they decide to shop elsewhere. Still others may not be able to find the right product, or may want to see the shipping charges before committing to a purchase. The results of improving conversion rate are shown in Table 13.1.

These problems are all amenable to usability testing. A usability test can be carried out for literally a fraction – 1 per cent – of the advertising campaign described above and once the problems are fixed, conversion rates usually more than double. In our example, this would result in an acquisition cost of just £2.50 per new customer.

The moral of this story is a simple one. The success of your web site is not determined by a rapidly-changing hit counter, or 'number of eyeballs', or even by the number of page impressions. It is defined by your conversion

Table 13.1 The relationship between conversion rate and sales for a site averaging 100,000 visitors making an average purchase of £15

Conversion rate	*Sales*
2%	£30,000
3%	£45,000
4%	£60,000
5%	£75,000

rate. And to improve your conversion rate you need to observe customers to find out where they stumble and fail to achieve their goals.

Fulfilment

By this stage of the book, it should be apparent that usability is about more than screen design, or even interaction design: usability addresses all of your business processes. This includes issues such as fulfilment and customer retention. It makes no sense to have an easy-to-use web site if week-long delays are introduced at the delivery stage. So when assessing the usability of your site, be sure to include measures of these 'beyond-the-interface' criteria.

But fulfilment is about more than just making sure the product arrives punctually. What customers are looking for is some feedback that their order has been placed and is being dealt with. On the Internet, you can never be sure if a company is being run by a teenager in his bedroom or a potential conglomerate with millions of dollars in venture capital. So customers want to know if their order is being dealt with seriously or has disappeared into the ether.

There are a number of stages in the fulfilment process. Immediately the order has been placed, the web site should display a page that acknowledges the order, confirms the content of the order and when it is expected to be sent, and provides any necessary tracking information (such as an order number or a URL). At the same time, the site should generate an e-mail that repeats this information and also includes contact information should the customer need to make changes (this should contain an e-mail address and a telephone number as a minimum). If there are any delays to the order, inform the customer immediately, and certainly before the shipping date. Once the order has been despatched, send an e-mail to the customer confirming this. Finally, a few days after the order should have arrived, send a further e-mail to the customer just to check that the goods did arrive. The customer only needs to respond to this e-mail if there has been a problem – and if there has been a problem you want to be informed quickly so that you can resolve it.

Of course, no-one wants to be bombarded by e-mails, but experience shows that it is difficult to provide *too much* feedback of this sort in the fulfilment process. In some instances, this handholding is more important than simply getting the goods out quickly. It is a bit like taking the exit from a motorway in the UK: one mile before the slip road, you see the first signpost showing where the exit will take you; this is repeated half-a-mile from the slip road and then just before the slip road; finally, countdown markers (III, II, I) lead you onto the slip road itself. Drivers don't complain about this; and your customers won't complain if you keep them fully informed about the progress of their order.

Usage rate by registered customers

Do your customers come back once a day? Once a week? Once a month? And when they come back, how long do they stay for? What parts of your site are of most interest to them?

How many customers buy from you several times? How many buy from you once but never return again? If customers buy once only, you will need to generate a lot of traffic to your site to maintain sales. In contrast, it is much easier to sell to an existing customer than generate a new one.

To measure customer retention, you will need a process to keep track of customers. Ideally, as soon as someone comes to your site you will know if they are a new customer or a repeat visitor. If this is a repeat visitor, you can customise the site and perhaps point the customer to an item that will complement the last item that he or she purchased. To do this, you will need to track customers either by using cookies or by a registration process.

Tracking customer comments

Customer comments take three main forms:

- e-mail/feedback form comments;
- surveys;
- helpdesk reports.

E-mail/feedback form comments

For every customer that complains about a feature at your web site, there are dozens more who experience the problem but do not take the time to tell you. This means that customer complaints are a rich source of data – rather than feel bad about them you should encourage customers to tell you how to improve. This means that simply putting a 'Contact us' e-mail link on your site is insufficient – if you want to actively solicit comments you should provide a self-explanatory link (e.g. 'Comment on our web site') and a proforma.

To get as many comments as possible, consider rewarding people who take the time to make a comment. For example, you could have a monthly draw in which you select a comment at random, e-mail the author and offer the customer a small gift (such as a sample of one of your products).

Most importantly, be sure to reply to everyone who comments on your site. It is easy to generate an automatic e-mail immediately the comment has been submitted and tell the customer when he or she can expect a proper reply: 'Thank you for taking the time to comment on our web site. One of our team will get back to you within 24 hours'. Be sure to contact customers within the allotted time: if your expected turnaround time is two to three days, it is better to say that you will reply 'within a week' and impress customers with your efficiency, rather than say you will reply 'within 2 days' and then disappoint the customer with a tardy e-mail.

Your reply should acknowledge the customer's problem, provide a solution and thank them for their comment.

Surveys

Surveys were discussed in Chapter 6. Although surveys have weaknesses when used to uncover usability problems, they are a useful data collection method when asking customers about factual data. For example, you could validly ask customers which parts of the site they are most likely to revisit, or what related products they would like you to stock, or if they managed to find what they were looking for. Answers to these questions can be interpreted more validly than subjective questions such as 'Did you find our web site easy to use?'

Helpdesk reports

Helpdesk data can be invaluable to identify critical usability problems with your site. People call a helpdesk because they desperately want to complete a task but have reached an obstacle. Usability specialists describe this as 'critical incident analysis' and solving these problems will fix any 'show stopper' usability problems at your site. For example, if a customer rings up to confirm that his order went through, this indicates that better feedback is required after the purchase has been made.

A second useful feature of the helpdesk is to show developers what life is like as a user of their web site. Developers should be encouraged to spend a day on the helpdesk, either fielding questions or (as a minimum) listening in to the questions that are asked.

E-commerce evolution

As the web evolves, the context of use will change: customers become more sophisticated, the technical environment changes and customers' tasks may change. Your site needs to adapt to this new environment.

Customers change

Customers' expectations change as they use more of the web. A feature may not have worked a year ago because customers did not want to use it for fear of looking stupid. A year later, that same customer has had the feature demonstrated by a colleague on a different web site and now expects to find the feature on your site. For example, a competitor site may offer real-time sales assistance in a 'chat' window. If this becomes an accepted and useful feature at other sites, your site may need to offer it too.

However, as with all new technologies, be sure to introduce changes in a user-centred manner. Some changes, like gratuitous animation, may have an initial 'wow' effect but can soon become passé and annoying.

To take a specific example demonstrating the way customers change over time, consider the perceived risk of fraud on the web. Risk of fraud is the oft-cited reason given by most people who do not make purchases on the web. Customers that are used to purchasing from web sites are aware that they are just as likely to experience fraud in the local high street as they are to experience fraud on the web. So the real fear of fraud comes from customers who are new to e-commerce. To address this concern, some sites also include 'fraud' agreements, and these are much more useful (and succinct). The principle here is that should the customer experience any loss as a result of using your site, you will reimburse the customer.

A related issue is confidentiality: customers are not always clear if their personal details will remain private or if they will be passed on to a third party (only to re-surface in a junk e-mail). Many sites include a 'privacy policy' that describes how the site will use the personal data that it collects, and as a minimum this should comply with legislation, such as the Data Protection Act in the UK. But, like a legal 'terms and conditions' agreement shown on entering a site, these are rarely read. The best way to address confidentiality concerns is by passing on some confidence to the customer as your site is used. A site that displays advertisements in a pop-up window each time a link is chosen conveys the impression that the business model behind this site is advertising-driven. Under this condition, customers will be reluctant to reveal any personal details about themselves because of the risk that it will be passed on to other companies. The best advice is to behave ethically and demonstrate that principle in the design of the site.

Environments change

Browsers evolve. When an upgrade appears, you need to check that your site still works. Finely honed HTML code that looks great in the last version of Browser X may break when viewed in a newer browser that parses code more strictly.

Technology evolves. New kinds of devices enter the field, such as WAP and mobile devices. If your customers are likely to use these technologies to access your site, you will need to support them. Predicting the future is, of course, difficult, but you can control this risk by structuring your site correctly. This will ensure that the main changes will be to the visual design of the site and in making the content more succinct.

Tasks change

As the web evolves, your customers' tasks and requirements will evolve too. Customers will expect to do more and more with your site. For example, if you run an on-line bank it may be fine in your first release to offer access to statements and a bill-paying function. But later, as your customers become more comfortable with the technology, they will want more: for example, to get an overall look at their finances. They may have accounts at different financial institutions and want you to aggregate their accounts so they can see them all in one view.

'Listening in' on customers' search queries will help provide you with an early indicator of changing customer tasks. There is some evidence to suggest that customers are more successful at finding what they want on a web site when they have to use the navigational/menu items than when they use search. One reason for this may be the parlous state of most search engines.

Imagine that you entered a bricks-and-mortar store intending to buy a sweater and asked a shop assistant where to go. If shop assistants were like most search engines, the conversation would go something like this:

You:	Hello. I'd like to buy a sweater. Can you tell me where I will find one?
Assistant:	Cannot find any instances of search term 'sweater'.
You:	OK. How about jumpers?
Assistant:	Cannot find any instances of search term 'jumpers'.
You:	Mmmm. Hold on a second. I know: Jersey!
Assistant:	Cannot find any instances of search term 'jersey'.
You:	One last try: Pullover?
Assistant:	Menswear, third floor.

In reality, you would probably leave a real store long before you tried the 'correct' term, but the point is that a good proportion of your customers will search your site using synonyms for your products and not the term that you use in your sales brochure.

Some of these terms can be added using a thesaurus. The synonyms can then be added to the search engine so that when a customer searches for 'sweater' the search engine translates this into 'pullover'. But this is only part of the solution. You should also look at your access logs and scan the search terms that customers use: this will help identify new or fashionable names for the item and will also reveal common misspellings of terms.

A second benefit of monitoring search terms is that you also get to see the type of content that your customers want and expect to find at your site but perhaps you do not provide. This can be used as a marketing tool to identify the types of content or products that you could usefully add to the site.

Process evolution

As a final step, be sure to review the usability processes that you used on the project. Hold a post-implementation review meeting to identify those usability practices that worked well and those that could be improved. You may find that re-taking the test in Chapter 2 will help you identify some of these areas. Then discuss ways of improving your design, development and implementation process so that you can capitalise on the strengths and address any weaknesses.

Summary

- Tracking the performance of your site means more than simple site maintenance.
- There are several key statistics to track: conversion rate, fulfilment and customer retention, and usage rate by registered customers.
- Other items to pay attention to include: monitoring search terms, customer comments, surveys and data from the helpdesk.
- Web sites can never be static: the context of use changes. Be sure to pay continuous attention to changes in the customer base, changes in the technical environment and in the tasks that customers want to complete.
- Hold a post-implementation meeting to discuss the effectiveness of usability processes and identify areas of improvement.

Further reading and weblinks

Books and articles

Bias, R.G. and Mayhew, D.J. (1994) *Cost-Justifying Usability*. San Diego, CA: Academic Press.

Dillman, D.A. (1999) *Mail and Self-administered Surveys*. New York: John Wiley & Sons.

Pogue, D. (ed.) (1998) *Tales from the Tech Line: Hilarious Strange-But-True Stories from the Computer Industry's Technical Support Hotlines*. New York: Berkley Publ. Group.

Web pages

Nielsen, J. (1999) 'Usability as a barrier to entry'. http://www.useit.com/alertbox/991128.html. 'There are two important issues in Web marketing: 1. Getting people to your site in the first place: that's what the advertising budget is for. 2. Making people stay on your site and convert them from one-time visitors to regular users: that's what the usability budget is for.'

Rhodes, J. (1999) 'Web marketing research'. http://www.webword.com/interviews/lee.html. 'Web marketing research is about conducting research (usually statistical analyses) to understand web marketing issues'.

User Interface Engineering (2001) 'Users don't learn to search better'. http://world.std.com/~uieweb/Articles/not_learn_search.htm. 'When we watched 30 users trying to search various sites for content they were interested in, we noticed a peculiar phenomenon: the more times the users searched, the less likely they were to find what they wanted.'

14 What now?

This book has described a practical approach for ensuring usability in e-commerce projects. As with all practical approaches, the tools and techniques need customisation to become part of your day-to-day development process. Like many craft tools, customer-centred design techniques need to be tried out, lived with and adjusted to fit.

It is tempting to wait for your Next Big Project before applying these techniques for real. But if you do this, you will find yourself wrestling with both a new project and a new toolkit simultaneously. This approach may work, but like a pair of new hiking boots, it may be best to take the techniques out for a dry run before applying them in anger.

So the book ends with a list of suggested ways to try out the techniques on your own project next time you are back at work. Just pick one idea from this list and make a contract with yourself to try it out. That way you get to play with the techniques in a safe environment where there is no risk of failure. And who knows, your current project may even benefit from what you find out!

Some ideas for you to try

- Complete the stakeholder form for your current project and distribute it amongst your team. Do you share a common perception of the customers for the project?
- Complete a system profile for a current web site you are developing. Which characteristics do you find hard to complete? From where can you get this information?
- Ask somebody else on your project team to complete a customer profile for a site being developed. How does it compare with your own perception of the customer? How did you develop your views of the customer?
- Take a representative customer and ask them to carry out an important task with your site while 'thinking aloud'. Ask open questions but do not tell the user how to do the task. What will you change about the site based on this meeting?

- Make contact with some customers who recently e-mailed the webmaster or who completed a feedback form. Speak with them by telephone and ask some questions using the customer profile form as a checklist. Are they like you expected? Would you modify your customer profile in the light of your observations?
- Spend 15 minutes in the environment that your system will be used. Does the physical environment match what you expected?
- Identify the key objectives of the site you are working on. What key performance indicators could you use to track progress? What values might you place on them?
- Write down one important task that your site is supposed to support. Then ask someone new to the site to do it. Watch the person attempt the task but do not prompt or help. What surprised you?
- Spend 15 minutes carrying out an evaluation of just one of your screens using the heuristic evaluation technique. Which heuristics did you find most useful? What problems did you find and what implications does this have for your next design?
- Write down a frequent task that your site is supposed to support. Specify the step-by-step procedure that the ideal user will use to complete the task. Now carry out a cognitive walkthrough of your design (use Appendix 9 as a prompt). Are there any elements of the design that need to be revised?

Appendices

Appendix 1 Stakeholder analysis form

Stakeholder for Site Name | Author: | Date:

Stakeholder roles and main concerns	Management strategy	Size	Importance	Priority
Clients or sponsors How will I make money? What risks could prevent me getting a return on investment?				
Customers/users How will this help me do my job better, save time, make money or have fun?				
Shareholders/Investors Will this initiative increase or decrease the value of my shareholding?				
Testers How do I go about testing this site?				
Business analysts Will the developers include the results of my business modelling in their design?				
Technical support What are the main problems that customers are likely to have with this site?				
Legal experts How can I control the risk of copyright violations?				
Site designers How do I go about coding the design? Can I re-use code from other projects?				
Documentation experts How will I teach people to use this?				
Marketing experts Will this increase the company's brand value? Can I use it to collect CRM data?				
Competitors Will this initiative affect my market share? Can I use their ideas on my own site?				
Technology experts Does the site use the very latest technology? Are there any security concerns?				
Domain experts If I help design the site, am I doing myself out of a job? Will this de-skill me?				
Regulatory bodies in the industry How does this change the competitive environment?				
Representatives of trade associations How does the site compare with others in the industry? Which is the "best buy"?				

Size	L	M	H
H	3	2	1
M	4	3	2
L	5	4	3

Importance

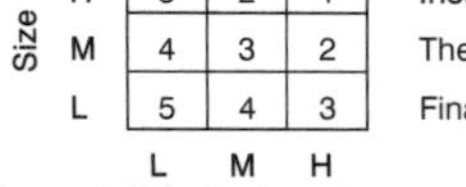

Instructions: Note the size of each stakeholder group as a proportion of all stakeholders (e.g. >75%=H, <25%=L).

Then note the importance of this stakeholder group to the success of the project (L, M or H).

Finally, use the table on the left to assign a priority to the stakeholder group.

Site mandate for site name Author: Date:

Characteristics	Description	Source and confidence
Project or site name		
Characteristics of e-commerce site Does a similar website already exist? Is this a new development? How many unique visitors will the site have? What likely conversion rate will you aim for (typical rates are 2%-5%)?		
Business model How will the site make money?		
Business and brand objectives What are the company's business objectives? What brand values should the website communicate?		
Key site objectives What are the key success factors for the site? What needs to happen for the site to be considered a success? What are the short- and long-term objectives of the site?		
Target market planned Is it a mass-market website or targeted at a narrowly defined customer group? Market sector; Market size; Geographical distribution of market.		
Value proposition What benefits will your customers get from this website?		
Likely functions What kinds of functions will the site probably have?		
Competitor websites Who is the market leader? Which specific websites will compete with this new development? If there are no competitor websites, what manual task is the website replacing?		

Appendix 3 Customer profile form

Customer profile for stakeholder group

Author:

Date:

Characteristics	Description	Source and confidence	Design implications
Personal/physical details Name; contact details (for later follow-up); native language; age; sex; likely problems with input device (mouse/keyboard etc.) or reading screen; physical limitations			
Job profile Why would they visit the site? How do they find websites (search engines, portals)? Expectations of site behaviour (conventional or unconventional?); Preferred navigation style (hierarchical/linear); Attitude to innovative UI styles (for example, Flash)			
Education Highest qualification achieved; key topics studied; developed applicable skills; reading level and comprehension; typing speed			
Domain knowledge Everyday concepts used when describing or using product; experience with current website; experience with competitor websites (which ones?); experience with websites that have similar functions; familiarity with websites in general; experience with specific applications, interfaces or operating systems (which ones?).			
Style preferences Intellectual styles (approach to problem solving); working styles; learning styles; preferred writing tone; attitudes to product; attitudes to IT; attitudes to brand; life values; discretion to use website.			
Concerns Current problems with carrying out these tasks; Key trends in this area; Aspects of task currently found unrewarding and would like to disappear/diminish; What they expressly don't want to see in a new website			
Wants Key enhancements they want to see; aspects of task found rewarding and would like to see more of.			

Appendix 4 Environment profile form

Environment profile for stakeholder group

Author:

Date:

Characteristics	Description	Source and confidence	Design implications
Physical environment What does the workspace look like (pictures if possible)? workspace size; space and furniture; light levels; noise levels; thermal environment; any temperature variations? visual environment; location relative to other people and equipment; location and availability of any equipment or information to support task; customer posture or position.			
Socio-cultural environment Task practices; team working; management structure; organisational aims (for B2B sites); industrial relations; communication structure; cultural issues; sharing; security/privacy; policy on training; assistance required or available; policy on IT; policy on Internet use; job flexibility; autonomy; hours of work; interruption; breaks; performance monitoring; performance feedback; pacing			
Technical environment *Hardware* Portable or fixed system; networked or standalone; frequency with which the system will be moved; type of input device (e.g. touchscreen) and output device (e.g. flat panel display); processor (e.g. Wintel). *Software* Custom or off-the-shelf software; type of operating system. *Reference Materials* Manuals; on-line help; "Quick Tips" card.			

Appendix 5 Task list form

Task list for stakeholder group　　　　Author:　　　　Date:

Task description	Frequency	Importance	Priority

Size	L	M	H
H	3	2	1
M	4	3	2
L	5	4	3

Importance

Appendix 6 Detailed task information form

Detailed task information for Task Description   Author: Date:

Characteristics	Description	Source and confidence	Design implications
Goal of task Why is this task performed? What higher-level task is it part of? What is the business context for this task? What are the criteria for success?			
People Which customer performs the task? What skills are required to be successful in the task? What other interactions with people are there? What communication with other people is needed to complete this task?			
Difficulty Is the task simple or difficult? What errors or problems do people have with the task? What mistakes are common? What are the consequences of errors?			
Time How frequently is the task performed? How long does the task take? Does the customer have any flexibility? Are there any time constraints on the task? Is any pacing imposed or necessary?			
Inputs, outputs and relationships Starting point or input; Where does the information for this task come from? What is the product (output) of the task? Where does the output go to? What information is required to successfully complete the task? How is this task related to other tasks?			
Results Results required (collect examples where possible); quality; how results are to be used.			

Appendix 7 Scenario form

Scenario Scenario name

Author: **Date:**

Scenario	Notes
Scenario checklist: Is it accurate and realistic? Is it specific and measurable? Does it describe a complete job (comprising integrated, not simple tasks)? Does it describe **what** the customer wants to do (not **how** the customer will do it)? Is the task "portable" to equivalent websites?	Use this column to justify assumptions you make about the way tasks are linked together, or design implications that occur to you as a result of the scenario.

Appendix 8 Key performance indicator form

Key performance indicator for site name Author: Date:

High-level objective

Context of use	
Who?	
Doing what?	
Under what circumstances?	

Measurement technique	
Scale/algorithm	
Data collection method	

Performance values			
Unacceptable range	Minimum range	Target range	Exceeds range

Appendix 9 Cognitive walkthrough form

Task description

Describe the task from the point of view of the first time customer. Include any assumptions about the state of the system (browser, website) when the customer begins the task.

Anticipated customers

Who will use this website? Do they have experience with similar websites or with earlier versions of this website? If so, which ones?

Customer's initial goals

Write down the goal that the customer is likely to have when starting the task.

Action Sequence

Make a numbered list of the atomic actions that the customer should perform to accomplish the task. Then transfer each step to the next worksheet.

Step 1:	Step 6:
Step 2:	Step 7:
Step 3:	Step 8:
Step 4:	Step 9:
Step 5:	Step 10:

Appendix 9 (continued) Cognitive walkthrough form

Step no.:

Will the customer realistically be trying to do this action?

Possible problems: *Notes*

- ❒ The customer will not expect to have to carry out this action at this point.
- ❒ It is not obvious that the action can be carried out at this point.
- ❒ The action is unnatural because customers are unlikely to have the knowledge or experience needed to do this action.
- ❒ Experience with other websites may mean that the customer is expecting to do a different action at this step.

Is the control (icon, button etc.) for the action visible?

Possible problems: *Notes*

- ❒ The control is hidden (e.g. context-sensitive, keyboard-only).
- ❒ The control is buried too deep within the menu or navigation system.
- ❒ The control for the action is non-standard or unintuitive.
- ❒ The control for the action does not have a label (and it needs one).

Is there a strong link between the control and the action?

Possible problems: *Notes*

- ❒ The label on the control is confusing (for example, an ambiguous or jargon term).
- ❒ There are other controls visible at this point that look like a better choice.
- ❒ There is a time-out at this step that does not allow the customer enough time to take the action.
- ❒ The action is physically difficult to execute.

Is feedback appropriate?

Possible problems: *Notes*

- ❒ Feedback is missing: customers may try to carry out the step again or give up on the task completely.
- ❒ Feedback is hidden: the feedback is easy to miss (e.g. hidden in a status bar).
- ❒ Feedback is inadequate: the feedback is too brief, poorly worded, inappropriate or ambiguous.
- ❒ Feedback is misleading: the website's response suggests that the task has been completed when it has not.
- ❒ Customers are not prompted to take the next step, or the prompt may lead them to take the wrong step.

Index